UP IN SMOKE

A Business Guide To Fire And Life Safety

By Robert Pessemier

Phoenix Publishing ◆ Redmond, Washington

Published by:
Phoenix Publishing
P.O. Box ~~3546~~ *15164*
~~Redmond,~~ Washington ~~98073-3546~~, USA
Seattle   *98115-0164*

Edited by:
Carol C. Schneider

Cover Design by:
Suze Woolf/Visible Images, Seattle, Washington

Excerpted Material:
One Minute Manager, page 39
by Kenneth Blanchard, Ph.D., and Spencer Johnson, M.D.
Copyright © 1981, 1982 by Blanchard Family Partnership and Candle Communications Corporation
All rights reserved
Reprinted by permission of William Morrow and Company

Library of Congress Cataloging-in-Publication Data

Pessemier, Robert
    Up In Smoke
    A Business Guide to Fire and Life Safety
    Includes Index
1. Fire Prevention — Methods
2. Life Safety — Methods
3. Business Management — Methods
I. Title
Library of Congress Catalog Card Number 86-62569
ISBN 0-940383-07-1 Softcover

Manufactured in the United States of America

For my family
Don, Bill, Phyllis, Bill, Rose, and Lindsay

ACKNOWLEDGMENT

I would like to thank the following people for helping make this book possible. Carol Schneider, John Ellis, N.H. Mains, Jeff Foster, Bob Leonard, Don Hamilton, Scott Dickson, Suze Woolf, Doug Plummer, Rod McClachlan, Chief Robert H. Ely, Chief Norm Angelo, Battalion Chief Jed Aldridge, Assistant Chief Marv Berg, Willie O'Neil, Tom Rutten, Ellisa Koch, Jan Salisbury, Chris Ostrom, Mike DeHart, Mary Ann Gillis, John Lee, Denise DiPietro, Patty Corbin, and the entire Kent Fire Department.

Special thanks to Pam for all her help and courage.

Table of Contents

DISCLAIMER

The author and publisher assume no responsibility or liability for any action taken by others as a result of the material presented in *Up In Smoke*. The author and publisher are not engaged in rendering legal or other professional services. If other expert assistance is required, the services of a competent professional should be sought.

The material presented here is the opinion of the author and does not represent that of any other person or entity.

PART ONE
WHERE TO START

1.
Why You Need This Guide

"Nothing is so firmly believed as that which is least known."
Francis Jefferey

Every year, $12 billion of American business resources go up in smoke. Some of this money is yours. You pay higher insurance premiums to cover fires caused by other business owners. If you have a fire, you pay out-of-pocket expenses because no insurance policy can cover all losses. You pay higher taxes to keep the firefighters and their shiny new trucks coming to the rescue. You may even end up in court with a liability suit. Seventy-five percent of this annual loss — nine billion dollars — can be saved by a simple and inexpensive form of investment — knowledge.

This book contains the knowledge you need to prevent a fire, keep you out of court, involve your employees in the business, and help you run your business more efficiently. It may also help reduce your insurance premiums. *Up In Smoke* will help you become a more informed and effective manager.

This book will inform you, educate you, and may even entertain you about fire and life safety. It includes information you have never seen before. This guide is the first of its kind, written especially for the business owner and his or her staff. It will give you ideas about how to start your own successful prevention program, and it will outline the steps to keep it running smoothly.

Up In Smoke is designed to do more for you than prevent fires. It will also help you avoid some of those nagging little problems associated with running a business, such as mechanical breakdowns and appliance failures. These and other headaches can be remedied by using the information in this guide.

Up In Smoke is designed to keep your typewriters typing, your machinery running, your employees feeling safe and working effectively, you out of court, and fire out of your business. The information in this guide can help you do all of this and more.

Up In Smoke is designed to keep your business profits where they belong ... in your business and in your pocket.

2.

How To Use This Guide

"To read without reflecting is like eating without digesting."
Edmund Burke

Use this guide as an information resource. Although it contains information you may not have seen before, don't worry. It's easy to read and understand, and it's especially easy to apply to any business situation. And it doesn't take much time to read either. Just reading this book once can reduce your chance of having a big problem.

Fire requires four elements: heat, fuel, air, and a human act or omission. If you eliminate any one of these, fire will not occur. This book is intended to give you the knowledge to eliminate the human act or omission that causes fire. Remember, knowledge is a cheap form of investment — for you and your business.

Up In Smoke is divided into three parts. Part One explains what led the United States to a fire loss record second only to Canada as the worst per capita of any industrialized nation. It also explains how this loss affects you and why you share in the cost, even if you don't have a fire.

Part Two is the meat of the matter. It spells out in plain English — not the language of codes — the fire and life safety hazards that may exist in your business. Part Two describes the hazards, tells where to find them, and explains how to eliminate them. This is the information you have always needed but never had.

The first nine sections of Part Two cover specific kinds of hazards. Any of these hazards can create a potential problem. The problem may be large and imminent or small and improbable. Nonetheless, it is important to know what they are and how to solve them. In addition, each section has a comprehensive, easy-to-use checklist for your personal use.

Even if you have taken all the preventive steps, it is still possible to have a fire. Section 10 of Part Two explains how to handle a fire if one does occur in your business. The last section in Part Two contains helpful information that explains how to deal with a fire inspection. It will make your inspection a painless and positive experience.

Part Three tells how to put all of this new knowledge to work for you and your business. It offers guidelines to start your own successful prevention program. In addition, it gives you ideas about how to involve your employees, increase productivity, and keep your business running smoothly. It also contains a resource section so you know who to call when you need more information or assistance.

As you identify the hazards and potential problems that exist in your business, you'll need to formulate a plan of attack. Basically, you can take three approaches to solve a problem: active, reactive, and passive. You may choose one approach, or adopt a combination of them. Keep these in mind as you read Part Two. The approach you select will depend on your particular business and whatever time and money constraints exist.

ACTIVE PREVENTION

An active approach is just that — active. This is an all-out effort to solve the problem. You do everything you can to make your business as safe as possible. Time, money, and effort are expended to educate all personnel, fix or replace defective machinery or building deficiencies, and do whatever else is necessary to bring your business to its peak of safety. Although this approach is the most effective way to reduce your risk of fire, injury, or a mechanical breakdown, it is also the most expensive.

An active approach can be applied to just one, a few, or all of the problems that exist. Obviously, this approach will depend on how much time, money, and effort is available.

REACTIVE PREVENTION

This approach is a less expensive way to solve a problem. You only solve those problems that prepare you for a fire. For example, you keep the sprinkler system inspected and maintained, the fire extinguishers charged and ready, and all exit paths and exit doors unobstructed. In other words, you do everything possible to prepare to react to a fire.

This method is similar to the active approach but it is applied only to certain kinds of problems. Ask yourself, What can I do now to minimize the effect a fire may have on my employees and my building? If you can't eliminate the risk (the active approach), do what you can to reduce the danger and the damage if one occurs. Prepare a good defense.

PASSIVE PREVENTION

This is the least expensive approach. It requires no money and little effort. All it takes is the time to read this guide. The key is awareness. The passive approach only requires you to be aware of the hazards that exist in your business.

However, for this approach to be completely successful, you must make sure all your employees understand the problem and its consequences if something goes wrong. For example, let's say you discover a defective electrical outlet. Tell your employees not to use it because it can cause a short circuit or an electrocution. It's that simple.

Awareness is the core of any prevention problem. If you don't at least pursue the passive approach, you don't have any approach. The possibility of a disaster showing up at your door will certainly increase if you don't recognize the problems. At the very least, you must be aware of the hazards to carry out any kind of prevention program. The passive approach is the minimum effort that underlies all others.

Although the passive approach is the cheapest investment for your employees and your business, it is probably the most cost-effective. Seventy-five percent of all fire and life safety problems can be prevented just by being more informed and aware of the hazards. People cause fire and life safety problems — people that are uninformed and unaware. These are the individuals that blew nine billion dollars last year, and some of that money was yours.

The more you and your people know about hazards, which face all businesses, the more you can reduce the chance of ever experiencing a loss. Encourage your entire staff to read this guide, or at least Part Two. It provides the information to help you eliminate the human act or omission that causes fires. This book is an inexpensive investment that can prevent potentially disastrous losses. It may even help you reduce your insurance premiums.

Because every business and hazardous situation is different, you may want to use a combination of these approaches. Use whatever works best for your situation. If you become more aware of the potential problems, which is the basic intent of this book, you can greatly reduce the odds of your business going "up in smoke."

3.
Who Is Responsible
For The Fire And Life Safety Problem

"It isn't that they can't see the solution. It is that they can't see the problem."

G.K. Chesterton

The needless loss of nine billion dollars every year is a problem. This problem is preventable, because almost all fires are preventable. This problem is not isolated to the United States. Canada, Germany, Australia, and New Zealand also have huge dollar losses per capita due to fire. This may not come as any surprise, but Japan has one of the lowest dollar losses per capita in the world.

If we are to improve our fire loss record, and save ourselves a few billion dollars in the process, it will be helpful to understand how we earned this dismal record in the first place. Then, hopefully, we won't repeat our mistakes.

Four groups have contributed to this problem: the business community, the insurance industry, the fire service, and the government. Let's look at each of these.

THE BUSINESS COMMUNITY

"If you fail to plan, you plan to fail."

Unknown

It seems that most business people have a rather indifferent or apathetic attitude about fire and life safety. "Bah-humbug" may be a good way to put it. (He had a bad attitude and remember what happened to him.) Even without doing a thing to understand the scope of the problem or the hazards that exist in their businesses, most people believe, "It will never happen to me." Then when it does happen, they lament, "I never thought it would happen to me." Realize that every day, over 1,000 fires occur in businesses across the United States. It happens to businesses of all kinds, all sizes, and in all fifty states.

This attitude, or rather this misconception, does not seem to be entirely intentional. It is born from a lack of understanding and information. Very few colleges, universities, or other learning programs include any training or information on fire and life safety. The topic doesn't get much press. It's hard to find solutions to problems you don't even know exist. The less you know about fire and life safety, the more you increase your risk.

This situation is like walking into a mine field and not realizing it. You probably won't realize it until you step on a mine. It then becomes abundantly clear that you have to watch your step—if it's not too late. The same is true for fire and life safety. If you don't know what the hazards are, where they exist, and how to correct them, you may find yourself in the midst of a fire.

Remember too, that ignorance is no excuse in the eyes of our judicial system. Liability suits are extremely expensive. The practices in this book will be your road map through the mine field of fire and life safety hazards.

The business owner's misconception, "It will never happen to me," keeps the community's head in the sand like an ostrich. And you know what part of the anatomy is sticking up. This is not the posture for an informed and effective manager.

Another common misconception is, "My insurance company will cover everything." This is rarely the case. The losses from a fire can be expensive and complicated.

Fire causes two kinds of losses: direct and indirect. Direct loss is usually straightforward. It is the result of damage or destruction to your building or its contents. Indirect loss is incurred from lost income, moving expenses, liability suits, retraining employees, rent or lease payments that must be made even if you are temporarily out-of-business, and any other extra expenses.

The indirect loss can exceed the direct loss. Three of every four dollars in an average fire loss is attributed to indirect loss. In the United States alone, that loss means nine billion dollars. Let's take a closer look at these indirect losses.

A major indirect loss occurs when your business is interrupted. Any fire, large or small, will cause an interruption. Once the fire department arrives, it has legal control of your business. The fire department will not release your business until it has established the cause of the fire and determined that the building is safe to occupy. This may take one hour, one day, or one year. In the worst case, the fire department can determine that you may never use the building again. Whether you will be allowed to return depends on the extent of the damage.

Even if the fire department does release your building, you may have to stay closed until you can replace machinery or make necessary repairs. You may still be paying rent and other operating costs but you will not be in business. In other words, you will be paying but not producing. You need specific business interruption insurance to be reimbursed for this loss. You will also have to keep very specific records to justify your claim.

Another form of business interruption can occur even if you don't have a fire. The fire department has the legal right and responsibility to close your business if it discovers significant fire or life safety hazards during an inspection. If the inspectors believe you or your employees are in danger, they will order your business closed until the situation is corrected. Unfortunately, most insurance policies don't cover this.

If you are not being inspected by your fire department now, you probably will be in the near future. Whether your business is shut down by fire or the fire department, it can be a very inconvenient and costly form of business interruption.

Other indirect losses can result from lost rental income or lease payments if you own rental property damaged or destroyed by fire.

You can also lose more time and money while you do all those necessary tasks just to get back in business. You won't be working to make a living — you will be working to get back to making a living.

In addition, if your business is totally or even partially closed for any period of time, some of your employees may have to leave to find other work. They need to make a living too. If they don't come back when you need them, you will have to find and train new people. This will also cost you time and money.

Many more situations exist that can reduce your income or cut your profits if a fire occurs, including:

- loss of customers
- lost return on capital investment
- loss of credit standing
- seizure of insurance payments by creditors
- overtime wages (fires can take extra work to get back in business or clean up)
- moving costs
- temporary office space costs
- demolition costs

What's more, a liability suit can really make your day. If you have a fire and someone is hurt or killed because an exit was blocked, you will be invited to court. Or let's say you have some faulty wiring, a customer unwittingly touches it, and presto — you are in court again. It will cost you a bundle, both mentally and financially, and you won't be happy.

Talk with your insurance agent to find out exactly what is, and is not, covered by your policy. Finding out after a fire can be very expensive and inconvenient.

Your insurance company will be glad to review your policy with you. Insurance companies are not trying to hide anything. They are trying to do their best to cover you for most reasonable losses. But they do work under legal and governmental restrictions and regulations. It is a complicated business and it shows by how the policies are written; the policies are in a different language.

Finally, one of the most widely discussed misconceptions is, "I have no control over my insurance premium." This may surprise you, but you may be eligible to significantly reduce your premium.

Insurance companies expect you to follow code. But if you

demonstrate an active awareness and involvement above basic code requirements for fire prevention and life safety, most companies will reward your efforts with premium reductions.

If you increase your individual efforts to protect your business against a fire or life safety claim, your insurance company may reduce your premium by as much as 25 percent.

Part Two and Part Three help you develop your own fire prevention and life safety program to eliminate the hazards that exist in your business. This kind of active involvement will not only reduce your risk of loss, but it may also lead to premium reductions.

If you are concerned about rising insurance costs, talk to your insurance agent or company. Find out if you are eligible for these reductions and what you need to do to qualify.

The $12 billion annual fire loss is more than the combined profits of the insurance and finance industries in 1984. It seems inconceivable that the business community will allow this waste to continue, especially when the cause can be so easily and effectively eliminated simply by gaining a little more knowledge.

The essential point is this: you will lose income and profits if you have a fire. How much you lose, or if you lose at all, depends on the steps you have taken to protect your business from the risk of fire. The more uninformed and indifferent you are about fire prevention, the greater your risk becomes.

THE INSURANCE INDUSTRY

"Where is the knowledge we have lost in information?"
T.S. Eliot

Benjamin Franklin was keen on knowledge. He also founded The Philadelphia Contributorship for the Insurance of Houses from Loss by Fire in 1752. This was the first successful fire insurance company in the United States, and it is still open for business today. We have needed insurance for a long time. Mr. Franklin was also one of America's first firefighters. He realized the benefits of both services.

Ironically, as much as most people look upon the fire service with regard, they look upon the insurance industry with doubt and suspicion. This is not entirely deserved for such a complex and necessary service.

Unlike the fire service, whose money comes from taxes, the insurance industry's survival depends on its ability to turn a profit in a very risky line of business. Your insurance company invests your premium dollars and hopes the return is large enough to pay the bills, the claims, the agents' commissions, and a large enough dividend to keep the shareholders happy. This is an oversimplification but you can see that these companies are, in a way, at the mercy of their customers. This is especially true during a time when it is difficult to make profitable investments. If the return on investments is poor, and the losses and operating costs are high, an insurance company could find itself in big trouble.

During the early 1980s, insurance companies concentrated on attracting customers and investing their premium monies. And with a 15 to 25 percent return on that money, who wouldn't? But little effort has been spent to educate policyholders about how they can individually reduce their risk of fire and life safety losses. In the scramble to make a profit in this highly competitive business, insurance companies have let nine billion dollars slip through the cracks.

Everyone needs to be more aware of the fire and life safety hazards that come with operating a business. Insurance companies could do more to prevent fire loss claims if they put more effort into education. If you know where you may have a problem, you can take steps to solve it.

You may be asking yourself, Why should I be bothered with this? After all, you are paying an insurance company to assume some of the financial risks involved in operating a business. This is just the point. Insurance companies only assume part of the risk, not all of it. They are in business too. Insurance companies are judicious about selecting the risks they will assume. This may be one reason their policies are so hard to decipher. They want some breathing room. They are not trying to deceive you; they are just trying to stay in business. Remember the analogy of the mine field. If you have a fire loss, the insurance company may pay most of the bill, but you and your business are still going to be hurt.

You will also be hurt even if someone else "steps on a mine." If another business covered by your company files a claim, every business with that company may end up paying for that mistake. If the losses are too big, the insurance company may have to raise everyone's rates. This can get very expensive. Some insurance premiums have risen 400 percent in the last few years.

A fire or life safety loss is expensive for you, and a claim is expensive for the insurance company. Most businesses need insurance, and most insurance companies need you. It is in everyone's best interest to prevent loss. Policyholders and insurance companies need to get better at preventing loss.

The insurance industry has contributed to America's fire loss problem because it has not supplied you with the necessary tools to help you reduce your risk. Even though most companies now offer premium reductions for businesses that actively protect themselves against loss, business owners need to understand the mechanics of how to prevent loss. This will help business owners reduce their risk and potentially, their premium. Education and information are tools that can stop this nine billion dollar waste of money and resources.

Insurance companies have the statistics and information to help business owners prevent loss. If insurance companies want to help themselves and their profit margins, they must give you the information and support you need to help yourself.

THE FIRE SERVICE

"Two-hundred years of tradition unhampered by progress."
Fire Service Adage

Your insurance premiums pay for financial protection. Part of your taxes pay for the fire service to provide physical protection for you and your property.

So why is it that the second worst fire loss record per capita belongs to the world's most powerful and technologically advanced nation? Only Canada has a worse record.

A large part of the responsibility falls on the shoulders of the fire service. This arm of local government is charged with the responsibility to protect your life and property from fire and other disasters. The fire service is clearly an essential, honorable, sometimes heroic, and dangerous profession. However, its services are usually provided too late — after a fire occurs.

Historically, the fire service has been characterized as a reactive rather than a pro-active organization. If the fire service responds to an emergency at your business, it will try to minimize the damage and remove the risk. For too long, the fire service has been content reacting to situations that have already put you at risk, rather than doing everything possible to prevent them.

The fire service seems to think that responding to more fires faster, with better equipment, and more manpower is a sign of what a good job it is doing. This has perpetuated the rising costs of maintaining our fire departments. The cost is passed along to you through taxes. As more fires occur and the losses mount, fire service management looks for more opportunities to increase the staff and buy more equipment. The local government, seeing the increased losses, tries to act in your best interest and approves these requests. You pay more taxes. Without an active rearrangement of money and manpower to fire prevention activities, a self-perpetuating cycle will continue to cost you more money.

The fire service has never fully realized that its first and foremost responsibility is to prevent fires. Instead, it reacts to the situations it is in the best position to prevent. Although the attitude of the fire service is changing and progress is being made toward more emphasis on fire prevention, it is very little and very late.

The fire service has the knowledge, facilities, and personnel to assist you before you and your property are put at risk. Prevention is the only true form of protection from the risk of fire.

The most effective and efficient way for a fire department to use its resources — which you pay for! — is to be committed to an energetic educational program on fire prevention and life safety. The service you pay for should be one of information, cooperation, assistance, and understanding. You are, after all, the customer. You pay for this service like any other. Some fire departments are doing an excellent job providing this service. Too many are not. If you realize now that you are not getting your money's worth or an acceptable level of service, let your fire department know.

THE GOVERNMENT

> *"The legitimate object of government is to do for a community of people whatever they need to have done, but cannot do at all in their separate and individual capacities."*
> *Abraham Lincoln*

Your local, state, and federal governments have also contributed to the annual fire loss. They have not done what needs to be done. They have not fully provided usable, practical information and assistance to help keep your business safe. Part of any government's objective is to provide public protection. Many governments have not reacted with sufficient strength to solve the problem of fire and life safety. The American public is not as well protected as it should be.

"What do you think the reaction would be if, every month, one fully loaded 747 took off from Washington (D.C.) and another took off from Los Angeles, and they collided somewhere over the Midwest, killing everyone aboard both planes. The official and public outcry would be deafening; the attention given this problem would fill newspapers and television screens across the country. Yet, that is approximately how many people we kill in fires every month in this country. And the outcry is far from deafening."

This tragic analogy was expressed by Mr. Robert H. Ely, president of the International Association of Fire Chiefs and chief of the Kirkland, Washington Fire Department. Mr. Ely was in Washington, D.C., in February 1986 soliciting support from Congress to appropriate funds for the Reauthorization of the Federal Fire Prevention and Control Act.

This act, which became law in 1974, was created to assist local governments with statistical data, research, training, and the development of educational programs on fire prevention and life safety.

The United States not only has the second largest fire loss per capita record in the world but also has the second largest fire death rate. Despite this, the act is being considered for major budget cuts and the elimination of some of its agencies because of pressure to reduce the federal budget deficit.

Mr. Ely also made a comparison between funding this cause and another sensitive budget item, national defense.

"It would cost the taxpayers $10 million dollars less to fund all federal fire prevention programs than to purchase one F-18 fighter aircraft."

This man certainly knows what he is talking about. The federal government has not done nearly enough to fund the programs that could be of such direct and enormous benefit to the public. The government will buy more F-18s but wants to eliminate funding for various agencies supported by the act, namely, the United States Fire Administration (USFA).

The USFA is a national clearing house for information. It gathers all those wonderful statistics about how we burn our-selves up every year. But the act has not been funded well enough to get this information to the public in a usable form. These agencies created by the act are doing the best they can with limited funds, but you get what you pay for.

Our local, county, and state governments seem to have a similar problem. Fire prevention programs, which could produce the greatest benefit to the public, are drastically under-funded and rarely produce any measurable results. These branches of government put your money — and it is your money they use! — into fire engines and firefighters, rather than into programs to help you live and work in a safer envi-ronment. They are reacting to fires rather than preventing them. These levels of government are not using your money in a cost-effective way.

The federal government may not be doing enough, but at least it is doing something. And as Mr. Ely pointed out,

"Federal funding for fire prevention programs is extremely cost-effective."

Fire prevention programs, whether local or federal, need much more support to effectively reduce the annual nine billion dollar waste.

The government has the apparatus in place to provide you with more information and assistance on fire and life safety. The various levels of government need to get this information to you in a usable form. Each level could do much more to inform and educate the public about fire and life safety hazards.

These hazards exist everywhere, regardless of your business or its location. Prevention takes education, and education takes funding. Federal funding of these educational programs is one of the most cost-effective expenditures to come down the road in a long time. Spending a few million dollars to save several billion is effective.

Unfortunately, the government has not realized the scope of the problem or the most effective way to solve it. This failing has contributed to the $12 billion dollar loss record. Most other countries that suffer huge losses probably share this failure to understand the problem and how to constructively solve it.

As Lincoln stated, we cannot do this in our separate and individual capacities. It is up to the government to help educate us about fire and life safety. The public must know more about prevention to reduce the risk of loss. The government has the capacity to provide this basic and essential service, but presently, not the desire.

4.

How To Solve It

"Leadership is ACTION, not position."
Donald H. McGannon

Once you know a problem exists, you can take steps to solve it. The same idea works for fire and life safety. The problems inherent in America's fire loss record can be solved, and the time to start is now.

If the business community's problem is indifference because it lacks fire prevention information, then business owners need to be told how big the fire and death problem is, how much it costs them, and how they can help prevent it.

If the insurance industry's focus is inappropriate because it does not give you the practical information you need to prevent loss, then the industry needs to shift its focus to prevention. The industry must give you that information and make it worthwhile to use it.

If the fire service does not provide the service it is being paid to deliver because it does not assist and encourage the business community in prevention efforts, then the fire service needs to become more pro-active. The fire service should focus on fire prevention and offer its knowledge in a positive way for businesses to use.

If the fire service does not provide the service it is being paid to deliver because it does not assist and encourage the business community in prevention efforts, then the fire service needs to become more pro-active. The fire service should focus on fire prevention and offer its knowledge in a positive way for businesses to use.

If the government does not provide for the public good because it does not support all agencies and programs that can reduce the annual fire loss, then the government needs someone to explain simple economics. An investment of a few million dollars to save several billion is a smart investment. Any government should be proud to save its populace billions of dollars, not to mention thousands of lives. Maybe the Navy can go without a few F-18s in the interest of public welfare.

An indifferent attitude, a lack of awareness, communication and practical information, and feeble governmental support blend together to give this country, and others like it, dismal fire losses every year. We are ineffective in our approach to fire and life safety. Everyone has a hand in it and everyone is affected. The most cost-effective way to start solving the problem right now is to give you the information in this book.

As a business owner you need information to make intelligent decisions. How else can you manage your business affairs, or your risks? This book will give you easy-to-use information to help evaluate how safe your business really is.

Consider the risks, be prepared for their consequences, and make your own decision how to manage them. Don't risk more than you can afford to lose. Take the time to read this book. It will help prevent your time and money from going "up in smoke."

PART TWO
HAZARDS – WHAT THEY ARE AND HOW
TO ELIMINATE THEM

1.
General Considerations

 This is the catch-all section. These hazards are relatively simple, but they can burn you out or drag you into court just as fast. These little beauties are worth some attention.

This section covers basic fire prevention tips for combustible materials, electrical systems, fire detection and alarm systems, and building construction.

Items that require more sophisticated fire prevention tactics are examined in later sections. These sections will introduce you to the hazards and equip you to effectively eliminate them.

COMBUSTIBLE MATERIAL

Don't accumulate any combustible material in an area where it can ignite and destroy your building. Keep this material in dumpsters or metal garbage cans. A pile of waste material is not only a fire hazard but also an eyesore for your customers. Keep your building neat and clean. This is a simple but effective fire prevention measure.

Ashes
Any ashes from a fireplace or furnace should be put in a metal garbage can for disposal. Ashes can insulate an ember that can stay hot for a week or more. When these ashes are stirred and dumped into the trash, the ember can ignite a combustible material such as paper or plastic. Don't put your ashes in a paper sack and leave it on your doorstep, back porch, or anywhere except in a metal trash can with a lid. This applies to your home as well as your business.

Cigarettes
Cigarettes are a major cause of fires. They fall out of ash

trays and into crevices of furniture or walls only to smolder and ignite something hours later. If your employees or visitors smoke in the building, give them the proper ash tray.

Use ash trays that hold cigarettes in the middle of the tray, not on the edge of it. Then deposit all ashes and cigarette butts in a metal garbage can for disposal.

Oil

Keep all oily materials off the floor. A layer of oil on a floor is combustible. A little clean up will go a long way to prevent a fire.

Oily rags should be stored only in metal waste cans. These rags can spontaneously ignite, burn through a plastic can, and ignite an oily film on the floor or anything else that's handy.

Candles

If you use candles, find candleholders that have a glass bulb surrounding the flame and a large diameter base. The bulb will help prevent the flame from igniting anything nearby such as your hair, clothing, or window drapes. The large base will help prevent the candle from being knocked over and igniting something. In addition, it's not a good idea to set the mood with a kerosene lamp. It may turn ugly if someone knocks it over.

ELECTRICAL SYSTEMS

All electrical equipment generates heat. If your equipment is not maintained or used properly, it can generate too much heat — enough to ignite the wiring in the equipment and anything combustible nearby. In addition, accumulations of dust or oil can ignite. Keep your electrical equipment in good condition. A simple cleaning is a good practice for fire prevention.

Circuit Breakers

Unused electrical equipment should be unplugged or shut off at the breaker. Doing so will almost guarantee that this equipment will not start a fire.

All your electrical and heating appliances should be connected to a circuit breaker or some type of overcurrent protection. A surge of electricity or a short circuit can cause a fire in a flash.

Heating and Air Conditioning Systems

Heating systems produce just that — heat. Every system generates heat whether it is electric baseboard or gas with forced air. Keep all combustible material at least three feet from any heating equipment.

Lint and dust can accumulate and become an ignition source. Keep all heating and air conditioning equipment clean. This includes the cooling unit on your home or office refrigerator.

Call a local heating system expert to inspect your system on a regular basis. Not only is this a good fire prevention measure but it can also save you time and money. If you know how your system is performing and follow the inspector's advice, you can prolong the life of your system.

Electric Heating Blankets

Here's some advice to take home with you, or practice if you are in the home health care business. Never leave an electric blanket folded back on itself under other blankets when it's in use. This insulation can cause the blanket to overheat and ignite the bedding. When the blanket is not in use, make sure it's turned off.

FIRE DETECTION AND ALARM SYSTEMS

If you have a fire detection and alarm system, take good care of it and it will take good care of you. Continued false alarms are a real headache for your employees and customers. And, if your system fails to work, the fire will cause more damage. If you have a fire and someone is hurt, you may end up in court to explain why your system was not tested and inspected.

These systems give early warning of fire. They can save lives and property. Call a reputable company to inspect and test your system annually.

BUILDING CONSTRUCTION

Fire will try to find any and every way to spread. Leaving an open path for fire will only add to the damage. Keep all scuttles and trap doors, including garbage and laundry chutes, closed at all times. Also keep all walls and ceilings intact. Replace missing ceiling tiles and patch holes in the walls. These measures help prevent the fire from spreading and can save you a lot of money if you, or especially your neighbors, have a fire.

Post your building address so it is visible from the street. Let the firefighters find you when you need them. This is especially important for medical emergencies when seconds can mean the difference between life and death.

General Considerations Checklist

Date: _____   Name: _____

COMBUSTIBLE MATERIAL

_____ No combustible material accumulated on premises

Ashes

_____ All ashes contained in metal waste cans

Cigarettes

_____ All ash trays designed to hold cigarettes in the middle of the tray

Oil

_____ All floors cleaned thoroughly to remove oily film

_____ All oily rags stored in metal waste cans

Candles

_____ All candleholders equipped with a glass bulb and a large base

_____ All kerosene lamps banned from premises

ELECTRICAL SYSTEMS

_____ All electrical equipment maintained properly

_____ All electrical equipment cleaned regularly

Circuit Breakers

_____ All unused electrical equipment unplugged or shut off at breaker

_____ All equipment and heating appliances connected to a circuit breaker or similar overcurrent protection device

Heating and Air Conditioning Systems

_____ All combustibles stored at least three feet from any heating equipment

_____ All heating and air conditioning equipment cleaned regularly

_____ All electrical or heating equipment maintained free of dust or lint

_____ Heating system inspection completed regularly by a reputable company

Electric Heating Blankets

_____ All electric blankets unfolded when in use

FIRE DETECTION AND ALARM SYSTEMS

_____ Fire detection and alarm system tested and inspected annually by a reputable company

BUILDING CONSTRUCTION

_____ All scuttles, trap doors, laundry and garbage chutes closed

_____ All missing ceiling tiles replaced

_____ All ceiling holes patched

_____ All wall holes patched

_____ Building address visible from the street

2.
Electrical Distribution Systems

 This section describes potential problems associated with electrical panels, extension cords, junction boxes, wall switches and outlets, and electrical wiring.

More than 10 percent of all business fires are caused by electrical distribution systems. This 10 percent is the largest single cause of business fires. Because of the nature of the system, it can also cause the most damage to your business, both physically and financially. Let's examine how the system works and what you can do to avoid becoming a statistic.

Electrical distribution systems carry electricity through wires, extension cords, fuse boxes, junction boxes, circuit breakers, and multi-plug adapters. Obviously, these systems are vital to keep your business running smoothly. You can't do much business when the lights are out, the cash register is off, and the computer is down.

It is very important to maintain your system properly; otherwise, it can cost you time and money. Your business can be interrupted by something as simple as a tripped circuit breaker, or as serious as arcing or an electrocution. Arcing occurs when current jumps between two electrical contacts.

But how do you prevent what you can't see? Electricity doesn't drip out of wall plugs in bright purple hues when a problem arises. You may have a hazard in your business and not even realize it.

Let's use an analogy to evaluate the risk. Think of electricity as water and the electrical system as the water pipes. If you try to force too much water through a pipe, the pipe leaks. If you try to draw too much electricity through a wire or an extension cord, the cord overheats. This can occur if you use only one small-diameter cord for several office appliances such as your portable heater, Judy's typewriter, the lunchroom refrigerator, and the coffee pot.

Leaks in your water system cause something or someone to get wet. Leaks in your electrical system cause arcing, short circuits, and electrocution. Water causes inconvenience; electricity causes fires, injuries, interruptions, lost profits, and court appearances.

To avoid these problems, you need to know what conditions spell trouble and how to fix them. You can't see electricity; but you can see the electrical distribution system — wires, cords, plugs, boxes, and adaptors. The condition of these components indicates your risk of fire or life safety problems.

This section helps you identify some of the problems and offers some solutions. These prevention tactics apply to any system. This section deals only with the electrical distribution system and not with the appliances or equipment the system services. In other words, here we are concerned only with the "piping" and not what it services at the end of the line.

ELECTRICAL PANELS

Fuse boxes and circuit breaker boxes are two kinds of electrical panels. These panels are junction points for electricity. Your main power line is connected to the panel and then branches into several lines to supply wall plugs, lights, and everything else that runs on electricity. This panel is the starting point of your electrical distribution system.

Some panels have fuses; others have circuit breakers. The purpose is the same for either panel: both act as emergency on/off switches to protect you, your equipment, and your electrical system from receiving an overdose of electricity. For example, if your copy machine short circuits and you can't turn it off, the electricity continues to flow. This surplus electricity will damage the machine and perhaps cause a fire.

Another potential hazard is drawing too much electricity through the wiring. The wiring will overheat and ignite any combustibles nearby. In these hazardous situations, a fuse or circuit breaker can save the day. It will automatically operate (it "trips") to stop the flow of electricity.

Labeling

All boxes contain a main power shutoff switch and a series of fuses or breakers. Each fuse or circuit breaker protects a certain line of electricity such as one that leads to the heating system, the lights, or other equipment. Label each fuse or circuit breaker with the system or equipment it protects. If your box design makes labeling difficult, draw your own diagram and post it in or near the panel. If everything is clearly identified, you won't waste time fumbling around for the right switch in an emergency.

Location

Keep the panel visible and accessible. This may not be aesthetically attractive, but it is functional. In an emergency, you or the firefighters may need to get to it in a hurry to protect someone from electrocution. You will waste precious time if the panel is hard to find or reach.

Firefighters will rearrange your merchandise if it's blocking the path to the panel, and it won't be left the way you want it. A three-foot clearance zone around your panel will keep it accessible and also protect it from falling merchandise or equipment. Signs that indicate the panel's location may also save you time and money.

Fuses

If your panel uses fuses, you need to keep a spare set on hand. A circuit breaker can be reset but a fuse must be replaced. Note the capacity of each fuse on your panel diagram. Capacity is usually measured in amperes (or "amps") and is stamped on the fuse. This is a measure of the amount of electricity that may safely pass through the fuse before the circuit breaker will trip.

The replacement fuse must be the same capacity as the one that tripped. For example, a five-amp fuse will trip long before a 50-amp fuse. If you put a five-amp fuse in a 50-amp slot, it will continually trip and shut down that line. On the other hand, if you put a 50-amp fuse in a five-amp slot, it will not provide the proper protection for the appliances or equipment on that line.

These capacities are important for the proper operation of your electrical system. They assure protection for you and your equipment. Make sure you have the right fuse in the right slot.

Circuit Breaker Slots

If your panel uses circuit breakers, it may have some empty breaker slots. These require dummy circuit breakers or protective covers. Even though the slot is not being used it must be protected from either dust, which causes short circuits, or someone's finger, which causes law suits.

EXTENSION CORDS

Extension cords are lengths of insulated wire that give you more reach from a wall plug to the equipment. They come in various lengths and diameters. The diameter of the cord indicates how much electricity the cord can carry.

An extension cord should have the same current capacity as the equipment it services. If you draw more current through the cord than it is capable of carrying, the cord will overheat and ignite whatever is nearby. If your extension cord does not list the current capacity, follow this rule of thumb. The extension cord should be of equal or greater diameter as the cord that connects it to the appliance.

Keep your extension cords in good working condition. This means no splices, damage, or deterioration. Don't tie knots in the cords. These conditions only invite a short circuit or an electrocution.

If your equipment cord is of the three-prong variety, use an extension cord with the same kind of plug; otherwise someone or something may get shocked. Your equipment has a grounded plug for a reason.

Plug all cords into wall sockets by the plug head. Cords plugged in by their exposed wires are dangerous. This is asking for trouble. It's short-circuit city.

Extension cords should not be substituted for permanent wiring. Cords that extend through walls or ceilings, under doors or carpets, or across floors are dangerous and frowned upon by the fire department. The holes in walls and ceilings provide escape routes for fire to penetrate faster and more extensively into other parts of the building. Cords that run underneath something become insulated and overheat. The other cords become worn or damaged and expose "hot" wires.

Install permanent wiring. Permanent wiring installed by an electrician and approved by an electrical inspector has a greater margin for safety than 25 extension cords.

For temporary electrical service, you may run an extension cord over the floor, but only if you have a protective strip covering the cords. This strip is usually plastic and has a slot running underneath its length that covers and protects the cord from being damaged or worn.

Extension cords should only service one piece of equipment unless a suitable multiple-plug adapter is used. The adapter should have a built-in circuit breaker and the same current capacity as the equipment it will service. The capacities must be the same; otherwise, the extension cord will either create a fire hazard or constantly trip the breaker. Adapters are available at most hardware stores and are definitely worth the small investment.

JUNCTION BOXES

Junction boxes can be round or square, and made of plastic or metal. They contain splices (or "taps") in the wiring where two or more wires are joined together. A junction box covers and protects the splices or taps. The box protects the wires from damage and accumulations of dust or debris, which can cause short circuits or arcing, and from inadvertently curious fingers.

One side of the box has a removable cover. This is the maintenance door and unfortunately, it is usually missing. If the cover is off, dust, metal shavings, or any other conductive material can collect in the box and cause arcing or short circuits. To do its job properly, this cover must be in place. Keep it on.

WALL SWITCHES AND OUTLETS

All wall switches and outlets should also have protective covers. This may seem like an insignificant detail, but the absence of these covers can cost you much more than the time it takes to put them in place. They not only protect the wiring but you as well. These covers are inexpensive and can be found in most hardware stores.

ELECTRICAL WIRING

Most of the wiring for your electrical system is hidden in the walls, floors, and ceilings. Any exposed sections should be protected from damage. Cars, forklifts, and push carts can accidentally smash wiring. Improperly stacked or falling merchandise can also do its fair share of damage. If the wiring is damaged, worn, or cut, you are in for problems. Your business can be interrupted by tripped circuits, arcing, fire, or electrocution.

You can avoid some of these problems by protecting your wiring. One solution is to run the wiring through metal conduit. This is inexpensive metal tubing and it provides good protection. Another alternative is to install a metal guardrail. A good material to use is three-quarter inch galvanized pipe.

If you have any loose and exposed wires, gather and attach them to something. It only makes sense to use non-conductive plastic loops or stringers, not metal staples.

Electrical Distribution Systems Checklist

Date: _____   Name: _____

ELECTRICAL PANELS

Labeling

_____ All fuses or circuit breakers labeled on panel diagram

_____ Diagram of fuses or breakers posted inside or near panel

Location

_____ Electrical panel visible and accessible

_____ Three-foot clearance zone maintained around panel

Fuses

_____ Extra fuses with proper capacities stored on premises

_____ All fuse capacities noted on panel diagram

_____ All fuses loaded into the proper slots

Circuit Breaker Slots

_____ All empty circuit breaker slots covered

EXTENSION CORDS

_____ All extension cords equipped with diameters equal or greater than the appliance cords they service

_____ All cords maintained in good condition

_____ All cord plugs mated with same kind of plug

_____ No cords with exposed wires plugged into sockets

_____ No cords substituted for permanent wiring

_____ No cords used under carpets or doors, or through walls or ceilings

_____ All temporary cords covered with slotted plastic floor strips

_____ All multi-plug adapters equipped with circuit breakers

_____ All multi-plug adapters rated with the same current capacity as the equipment they service

JUNCTION BOXES

_____ All splices and taps contained in junction boxes

_____ All covers installed on boxes

WALL SWITCHES AND OUTLETS

_____ All wall switches covered

_____ All outlets covered

ELECTRICAL WIRING

_____ All exposed wires protected from possible damage

_____ All loose and exposed wiring gathered and attached by non-conductive materials

3.
Commodities In Storage

 This section tells you how and where to store commodities. It includes general information about storage practices and specific information about how to store commodities near exits and fire protection equipment.

Every business stores materials. A commodity in storage is considered to be anything from a thousand boxes of merchandise in a warehouse to a box of receipts in the back broom closet. A commodity includes the product, the packing material, and the container or wrapping around it. For example, a metal product packed in styrofoam and packaged in a cardboard box, taken as a whole, is one commodity. It can be either one toaster in a small box or a dozen toasters in a large box.

How you store commodities will have a direct effect on the risk of fire and, if one occurs, how the fire department will operate.

Some part of every commodity will burn. All it needs is enough heat. Most commodities are either packaged in cardboard boxes, wrapped in plastic, or stuffed in paper bags. All these packing materials have low ignition temperatures. Commodities stored too close to a heat source will ignite.

But that's not all. Commodities make better doors than windows. If they are stored improperly, commodities can hamper fire department operations. Clearly, if the firefighters can't get to the sprinkler system or other vital areas, the fire will intensify and cause more damage. In addition, if someone is hurt or killed because an exit was blocked by commodities, you may have to make a court appearance. Good luck explaining your storage practices to a judge.

Keep these consequences in mind and store all commodities away from any heating or electrical equipment, fire protection equipment, or exits.

GENERAL STORAGE PRACTICES

Combustible commodities should not be stored in electrical rooms, mechanical rooms, or other equipment rooms where they can block access, cause damage, or start a fire. A box stacked too close to an electrical panel can ignite if the panel overheats or throws a spark.

In addition, combustible commodities should not be stored in attics or crawl spaces. Unless these areas have been specifically built for this purpose, the added weight will strain the supporting structures and the building may collapse soon after a fire begins. The commodities will also fuel a fire and make it difficult and time consuming to extinguish.

Also, keep commodities away from all motors, heating equipment, and anything else that gets hot.

Flammable liquids are extremely hazardous and suggestions for their storage are covered in Section Five.

STORAGE AND FIRE PROTECTION EQUIPMENT

As a general rule, commodities should be stored three or four feet from fire protection equipment. This space allows the firefighters immediate access to the equipment. It should also provide enough clearance to protect the equipment from falling commodities.

Store commodities at least three feet from any sprinkler heads or supply lines. Commodities stored or stacked too close to these areas will impede the flow of water out of the heads. Give the sprinkler heads enough room to function properly if a fire occurs. If a commodity falls, the clearance zone should prevent the commodity from hitting and breaking a sprinkler head. A broken head can cause a lot of water damage.

In addition, you should not store any commodities closer than three feet from the ceiling even if you don't have a sprinkler system. This three-foot gap will help the firefighters put the water where they need it.

Keep all storage at least three feet from the sprinkler riser. This is the control area for the sprinkler system and the firefighters may need to get to it in a hurry. The clearance zone also helps prevent damage from falling commodities. If one falls, hopefully it will miss the riser controls. Broken control valves can cause a lot of water damage if they are broken.

Sprinkler systems have at least one shutoff valve. The valve or valves are probably located opposite the riser on the outside wall of the building. If you store any commodities outdoors, don't block these shutoff valves. These valves should have a three-foot clearance zone from other objects. This zone will make the valves accessible to the firefighters and may also prevent the valves from being damaged by falling merchandise.

Your system may also have a fire department connection. This connection allows the firefighters to connect a fire engine to the sprinkler system to supply the system with more water. This connection is usually located on an outside wall of the building. Keep storage at least three feet from this connection.

Your fire extinguishers and hose stations should not be blocked by storage. Extinguishers can be moved, but stations cannot. If necessary, move extinguishers to a visible and accessible location, and remove all storage near the station area.

If your building is large you may have automatic fire doors. These doors will stop the spread of fire very effectively. Storage should not obstruct the operation of these doors.

STORAGE AND EXITS

It's important to keep all exits and exit paths clear of any storage. Exits are the last resort if a fire occurs. They are the only safe means of escape.

Keep all storage at least three feet from exit doors. Increase this distance if the building has oversized doors. This clearance zone will allow the doors to swing fully open.

Keep your main aisles four feet wide and unobstructed. This gives you and your employees a clear exit path.

If you have any doors that are permanently blocked by storage, especially large receiving doors, label them on the outside with four-inch letters. If the main entrances are blocked by fire, firefighters will try to gain entry through other doors or loading docks. A "This Door Blocked" sign will save them precious time.

Commodities In Storage Checklist

Date: _____ Name: _____

GENERAL STORAGE PRACTICES

_____ No commodities stored in electrical, mechanical, or equipment rooms

_____ No commodities stored in attics, crawl spaces, or under buildings

_____ All commodities stored a safe distance from motors, heating equipment, or other equipment that generates heat

_____ Flammable liquids — see Section Five

STORAGE AND FIRE PROTECTION EQUIPMENT

_____ All commodities stored three feet from sprinkler heads and supply lines

_____ All commodities stored three feet from ceiling

_____ All commodities stored three feet from sprinkler riser

_____ All commodities stored three feet from shutoff valves

_____ All commodities stored three feet from fire department connection

_____ All fire extinguishers and hose stations visible and accessible

_____ All fire doors unobstructed and operate freely

STORAGE AND EXITS

_____ All commodities stored three feet from exit doors

_____ All exit doors unobstructed and open fully

_____ All aisles unobstructed and maintained four-feet wide

_____ All blocked doors labeled "This Door Blocked"

4.
Waste Materials

 This section outlines the dangers of carelessly stored waste materials. It also offers helpful hints about how to safely store these materials until they can be permanently discarded.

Every business generates some kind of waste, and most of it is stored in or around your building until it can be disposed of altogether. The waste may be in the form of paper, wood, metal, cigarette butts, or generally anything you want to get rid of. This waste material can become a fire or safety hazard if it is accumulated in the wrong spot.

For example, old oily cleaning rags left in a pile can spontaneously generate enough heat to ignite. If this happens and they are not stored in a metal can, the rags will also ignite any combustibles nearby. Stacks of wood or paper are great targets for carelessly discarded cigarettes or sparks from nearby machinery.

Carelessly piled waste products are not only a dandy source of fire; they are also good obstacles for someone to trip over. Your employees or visitors may not have eyes in their feet, and consequently, you may have a needless injury on your hands.

Piles of waste can also reduce the effectiveness of the fire department. If you need the fire department and stacks of waste block the path, the firefighters will be slowed down and the fire will do more damage.

Waste material poses several hazards. It can start or fuel a fire, cause needless injury, and hinder fire department operations. Here are some suggestions for handling waste materials.

GENERAL STORAGE PRACTICES

Some oily rags can spontaneously ignite. Keep all oily rags

stored in a metal waste can with a lid. If they do ignite, the metal can will prevent the fire from igniting anything nearby.

Any combustible waste material stored inside your building should be kept in a metal garbage can or dumpster. Piles of wood or paper are excellent fuel for fire. Keep them in a large dumpster with the lid closed until they can be removed. If a fire does start, it will be contained in the dumpster. You can either buy a dumpster, or rent one from your local garbage disposal company.

Storing waste products in cans and dumpsters will also keep these combustibles off the floor and away from inattentive feet. If you can keep someone from tripping over this stuff, it may keep you out of court.

In addition, keep all waste material away from exit doors. If you have a fire, this material can block your only escape or impede fire department operations.

Outside storage of waste materials can also present problems. Remember, these materials are fuel sources and can ignite your building. This can be prevented by simply moving them a safe distance from the building.

If these combustible materials are stored in a dumpster, keep it at least 20 feet from all windows, overhangs or eaves, and combustible walls. Keep the lid closed; if this material burns, it will be less likely to ignite anything else.

In addition, anything stored outside should not block roadways or other access points. This will slow down the fire department.

Business offices and reception areas should use metal — not plastic — wastebaskets. All wastebaskets should be the same size. A metal wastebasket is non-combustible and can become a self-contained incinerator. For example, if a fire starts in one wastebasket, drop another wastebasket that is the same size into it. The fire will suffocate before it can spread.

Storage and disposal of cigarette butts and ashes, woodstove and fireplace ashes, and other miscellaneous waste products are discussed in Section One.

Waste Materials Checklist

Date: _____   Name: _____

GENERAL STORAGE PRACTICES

_____ All oily rags stored in metal containers with lids closed

_____ All waste materials stored in garbage cans or dumpsters with lids closed

_____ All waste materials stored off the floor

_____ All waste materials stored a safe distance from exit doors

_____ All outdoor storage of combustible waste materials stacked a safe distance from the building

_____ All waste materials stored in outdoor containers, cans, and dumpsters placed twenty feet from all walls, windows, and eaves

_____ All outdoor storage of waste materials stored clear of roads or accessways

_____ All office wastebaskets the same size and constructed of metal

_____ Cigarette butts, woodstove and fireplace ashes — see Section One

5.
Flammable Liquids

The storage and use of flammable liquids is a detailed and technical area of fire prevention. This section covers only the basic practices of flammable liquids storage and use. It explains why flammable liquids are dangerous, how they are classified, and how they should be properly stored.

This section also outlines the requirements for allowable quantities, flammable liquids cabinets, outdoor storage, spray painting rooms, and fire protection equipment. Contact your local fire authorities for more detailed recommendations.

Flammable liquids are everywhere. Hairspray, paint, copy machine ink, cleaning fluids, fingernail polish remover, and others are found at every business. Small quantities aren't usually a problem. But a bottle here and a gallon there can add up. Flammable liquids are usually the most volatile and dangerous substances found in any business. Be careful with any quanitity of these liquids. It will pay off in the long run.

The problem with flammable liquids is two-fold. First of all, the liquids are easily ignited. Second, they spread quickly over a large area if spilled. For example, a 50-gallon drum of gasoline, if upright, will burn and be self-contained. Knock over that same drum and the liquid will burn hundreds of square feet of material.

The most dangerous component of a flammable liquid is actually the liquid's vapor, not the liquid itself. When combined with air, this vapor can ignite and explode. The liquid's flash point indicates how dangerous it is. The flash point is the lowest temperature at which the vapors will ignite. Of the common flammable liquids, gasoline is the most dangerous. Its flash point is minus 45 degrees Fahrenheit.

It is not the liquids themselves that cause fires. But because the flash points of these liquids are so low, almost anything nearby can ignite them. Heating and electrical equipment, cigarettes, and even static electricity can be enough to set off these liquids. It is vital, then, to treat these substances with care and caution.

Flammable liquids come in many forms. Paints, solvents, fuel oil, and copy machine inks are examples of flammable liquids. The first order of business is to find out what kinds of flammable liquids you have and how much. These factors, along with each liquid's flash point, determine how each liquid should be stored.

All flammable liquids should be labeled "flammable" or "combustible" by the manufacturer. This may be all the information the manufacturer gives you but it is enough to indicate the substance deserves some respect.

Small quantities may be stored without too much special consideration. Store larger quantities in special lockers or rooms.

Four basic safety practices will help you prevent fires caused by flammable liquids.

Step One: Eliminate all ignition sources

Step Two: Keep the liquids in closed containers so the vapors can't escape

Step Three: Ventilate the storage area to prevent any vapors from accumulating

Step Four: Educate all employees about the hazards of flammable liquids and identify dangerous flammable liquids areas

---

You may decide to adopt one or all of these techniques. The measures you choose will depend on the nature of your business, the kind of hazards that exist, and how the liquids are used. But keep in mind that it only takes air, heat, fuel, and a human act or omission to start any fire.

This section describes how to determine some of the hazards of handling flammable liquids. Remember, these liquids are extremely dangerous. Consult your local fire prevention experts for more specific information on storage practices.

CLASSIFICATION

Flammable and combustible liquids are classified by their flash points and boiling points. The lower the points, the more dangerous the liquid. The classification indicates how dangerous a liquid is and therefore determines the technical aspects of its use and storage. Local interpretations and regulations vary. Call your fire department to discuss the situation in more detail.

Flammable liquids have a flash point below 100 degrees Fahrenheit. These Class I liquids are divided into three categories.

Class I-A: Flash point below 73 degrees Fahrenheit and boiling point below 100 degrees Fahrenheit
Class I-B: Flash point below 73 degrees Fahrenheit and boiling point above 100 degrees Fahrenheit
Class I-C: Flash point at or above 73 degrees Fahrenheit and boiling point below 100 degrees Fahrenheit

Class I liquids include acetone, benzene, gasoline, ethyl-alcohol, turpentine, white gas (stove fuel), paint remover, MEK solvent, JP-1 jet fuel, and some enamel paints.

Combustible liquids have a flash point at or above 100 degrees Fahrenheit and are labeled Class II or Class III liquids.

Class II: Flash point at or above 100 degrees Fahrenheit but below 140 degrees Fahrenheit

Class II liquids include fuel oil (diesel), JP-2 jet fuel, acrylic paint, Stoddard solvent, and kerosene.

Class III: Flash point at or above 200 degrees Fahrenheit

Class III liquids include motor oil, phenol, and bunker fuel.
If you have trouble finding the flash point of a liquid, call
your fire department, the manufacturer of the liquid, or the
chemistry department of your local high school or college for
more information.

GENERAL STORAGE PRACTICES

The fire department will only allow you to store a certain
amount of a flammable liquid in a certain kind of container.
These safety restrictions prevent you from storing too much
liquid in a breakable container or in one that may allow vapors
to escape. Check with your fire department for specific storage
requirements and limitations.

Don't store any Class I liquids in polyethylene containers.
The liquids will dissolve the container, leak, and expel dangerous vapors. Class II and III liquids may be stored in polyethylene containers but are subject to quantity limitations.

Flammable or combustible liquids should not be stored in
glass containers. These containers have a tendency to break
when dropped or knocked over. However, glass may be the
only way to store liquids that eat metal or plastic. If you use
glass containers, get the approval of your fire department.
When you call the fire department, be prepared to do some
explaining because you may know more about the substance
than the firefighter does.

The best way to store flammable and combustible liquids is
in approved safety cans. "Approved" means approved by your
local fire department and a testing laboratory. If the cans are
stamped with an Underwriters Laboratory (UL) or Factory
Mutual (FM) seal they will probably pass inspection. These
cans have tight-fitting, self-closing lids with pressure-relief
vents.

If you have large quantities of liquids, another good way to
store them is in 50-gallon drums. The drums should be in

good shape with tight-fitting lids and seals. If the drums have been used to store a different liquid, clean them properly before storing the new liquid.

If you transfer liquid from one drum to another, you need protection from static electrical discharge. This discharge can create a spark and ignite any vapors, blow you and your business sky-high, and generally ruin your day. Ground each drum to protect yourself from static electricity.

All small portable tanks or containers should have an automatic vent to regulate the internal pressure of the container. Although these vents create a potential hazard by allowing vapors to escape, the alternative is much more dangerous. A nearby heat source can send the internal pressure of a non-vented tank through the roof. If the tank blows, the fire will not be controllable. The automatic vents control and reduce these dangerous pressures. Check with your fire department to determine which containers are approved for your specific needs.

All flammable liquids containers should be labeled according to the U.S. Department of Transportation (DOT) standards. This labeling indicates what kind of liquids are involved so the firefighters can safely and effectively approach a fire. If the liquids are not labeled, it may take more time to extinguish the fire; consequently, the fire will cause more damage and that will cost you more money. Call your local DOT office for more assistance. Its employees will be glad to help you.

If you use solvents for cleaning, store them in approved solvent tanks. If a fire occurs in one of these tanks, the lid will close automatically to seal the tank before the fire gets out of hand. Use these tanks in a well-ventilated area. These tanks are definitely worth the price.

It is against the law to dump waste flammable liquids or any petroleum products on a street or in a ditch or sewer system. Disposing of this waste can be a problem. Your best bet may be a gas station with a waste motor oil collector. Otherwise, contact your local Environmental Protection Agency (EPA) office. If the EPA can't help you, it may know a private company that can.

Anytime you store flammable liquids you have a potential problem. The vapors may leak and find their way to an ignition source. Steps must be taken to eliminate, reduce, or control these ignition sources. This usually means that your heating and electrical systems must be kept in excellent condition. Contact your local electrical inspector or fire department for assistance and read Section Two about electrical distribution systems.

Oxidizing chemicals, or any other highly reactive chemicals, should not be stored within 50 feet of flammable liquids. Oxidizers have different reactions when exposed to various substances. Some oxidizers react with water and release oxygen, thereby intensifying a fire; some are explosive; and others actually decrease the ignition temperature of many flammable liquids. Any way you look at it, oxidizing chemicals don't belong near flammable liquids. They can make firefighting a real problem and cause needless damage.

Empty flammable liquids containers should be treated and handled as if they were full. Remember, it is the vapors that are explosive. Empty containers will have residual vapors unless they have been cleaned properly. If you change the contents of any flammable liquids tank, clean the tank properly before filling it with another liquid. Your local fire department will advise you how to clean the tank.

All flammable liquids containers should have their lids and seals in place. This prevents any vapors from escaping the tank.

Do not store flammable liquids near exits. These liquids can instantly block an exit during a fire. If someone is hurt because of this, you may be led into the courtroom.

Post warning signs to indicate the areas that contain flammable liquids. This will alert people to the immediate danger. A sign reading "Danger — Flammable Liquids" with three-inch-high red letters on a white background will do nicely. Your local fire department may have signs it will give you. Post these signs in visible places and make sure the letters are wide enough to be easily read.

OUTDOOR STORAGE PRACTICES

If you have a cache of flammable liquids outside, make sure it is stored a safe distance from the building. If space limitations force you to store liquids next to the building, install a fire wall along this portion of the building. Two layers of five-eighths-inch sheet rock on 2x4 studs will probably pass a fire department inspection. In addition, you may want to put a fence around the liquids to keep mischievous people out.

Don't stack containers. They make a dangerous mess if they fall over. If you store large quantities, you may want to construct a dike around the storage area.

If you have any 500-gallon or larger unused flammable or combustible liquids tanks sitting above or below ground, you may need the approval of your fire department to keep them. The fire department wants to make sure the above-ground tanks pose no threat of fire or explosion. Underground tanks pose two potential problems. First of all, if an improperly cleaned tank is accidentally punctured, an explosion can occur. Second, if a tank is forgotten and something is built on the land above it, an explosion or surface cave in can occur.

ALLOWABLE QUANTITIES

The fire department limits the amount and manner of storage of flammable liquids by the nature of the business. This section lists these parameters by business. Flammable liquids cabinets are described in detail later in this section.

Restaurants/Apartments/Hotels

You may store a maximum of 10 gallons of flammable and combustible liquids if they are used for the maintenance or operation of equipment and are stored in a safe place. If you have more than 10 gallons, you will have to store them in an approved flammable liquids room or cabinet. Get rid of any flammable liquids you are not using for the maintenance or operation of equipment.

These may seem like stiff requirements but you must realize that these businesses involve many people. The life safety hazard is extreme. You will have a tough time explaining yourself in court if you have a fire and someone is hurt because you stored more liquids than the code allows.

Schools and Offices

You may store any amount of flammable liquids needed for laboratory work, first aid treatment, or maintenance or operation of equipment. However, the fire department may require you to prove you need the stuff and know how to handle it. Generally, the requirements apply only to how the liquids must be stored.

If you have over 10 gallons, store it in a flammable liquids cabinet. You may store 120 gallons per cabinet with a maximum of three cabinets in any one area. If you have a larger quantity, store it in a flammable liquids room. Contact your local fire department or building code office for more details about these requirements.

Wholesale and Retail Stores/General Warehouses

The quantity limits for these businesses are technical and variable. Contact your local fire department for specific details.

Generally, the fire department does not restrict the amount of flammable liquids, only how they must be stored. You may have whatever amount is needed for normal and reasonable merchandising displays. However, do not store more than two gallons per square foot of storage space. This limit is reduced to one gallon per square foot if you store liquids above the first floor. Do not store any Class I-A flammable liquids in a basement. Basements are difficult for firefighters to get into and even more difficult for employees to escape.

Aerosol cans can be especially hazardous. Unless their contents and propellant are non-flammable, they are considered as Class I-A flammable liquids. These flammable liquids are considered especially dangerous in a fire because the containers can rupture and fly around like missiles.

The manner of storage is also important. Store all your flammable liquids close to the ground. Do not stack containers. The closer your flammable liquids containers are to the ground, the less likely they are to break if they fall over. If you use storage racks, make sure they are secured. Also, spread the containers over a large area. A lower concentration of these containers in one place will reduce the damage if a fire occurs.

Contact your local fire department for additional information about storage methods. Learn as much as you can about storage methods; exceptions exist for every rule.

Industrial Complexes

Storage and use requirements also vary by industry. Contact your local fire department for more specific details. Other valuable sources of information are industry newsletters and safety bulletins.

Service Stations

Don't store any Class I flammable liquids in a building with a basement or a work pit unless you can provide good ventilation. Flammable liquids vapors sink and can accumulate in these low spots. For other classes of flammable liquids, use the same storage guidelines as those established for retail stores. Keep all flammable liquids low and in a safe location.

If you do any welding, be careful. Welding is especially dangerous in the presence of flammable liquids. Ventilate the area well. If possible, install a blower or fan that is explosion-proof.

Your heating and electrical equipment can also become an ignition source. Keep it in excellent condition. Contact your electrical inspector for help.

Post "No Smoking" signs in visible areas. No one should smoke anywhere near flammable liquids. A carelessly tossed cigarette butt can be disastrous. Also, post signs instructing drivers to shut off their motors. A hot engine can be a great ignition source.

Your pumps should be protected from moving cars. Even a slow moving car can do extensive damage and cause a nice big gasoline leak. This wouldn't be good for business. Mount your pumps on a concrete island or install protective posts. Check with your local fire department for its requirements.

Also, install a master shutoff switch for your pumps. Label this switch so any customer can see it and get to it in an emergency.

FLAMMABLE LIQUIDS CABINETS

Contact your local fire protection equipment company for information about approved pre-built cabinets. If you want to build your own, follow these guidelines.

Metal Cabinets

Use steel that is about five one-hundredths of an inch thick. The entire cabinet should be double-walled with an one-inch airspace between each wall. Rivet or weld all joints to ensure a tight fit. Attach a self-closing door. Build a two- or three-inch watertight sill around the bottom of the cabinet to prevent leaks.

Wood Cabinets

The walls should be at least one-inch thick. Wood cabinets do not require an airspace. Secure all the corners and joints in two directions for added strength. Again, the door should be self-closing and fit tightly. The bottom of this cabinet should also have a two- or three-inch water-proof sill.

Cabinet Storage

Capacity and storage requirements are also important. The following guidelines should help reduce potential hazards.

You may build flammable liquids cabinets to any dimension, but you may not store more than 60 gallons of Class I or II liquids in any one cabinet. The maximum capacity of each cabinet is 120 gallons of flammable liquids. Space your cabinets at least 100 feet apart, and do not store more than three cabinets in one room.

SPRAY PAINTING

Most sprayed paint is extremely flammable. The paint is converted into a vapor spray that is easily ignited by sparks or anything hot. If you apply paint, stain, lacquer, or any other flammable liquid with a spray paint gun or similar device, follow these common-sense guidelines.

First of all, eliminate all dangerous heat sources that can touch off the vapors. Post "No Smoking" signs everywhere. Check your heating and electrical system regularly. Your electrical wiring should be explosion-proof or at least in excellent condition. Don't use any electric lamps, portable heaters, or any other equipment that generates heat in the same area where the painting is done. Remember, it doesn't take much to set off these vapors.

In addition, the spray painting area needs proper and adequate ventilation. This is crucial. The vapors must be removed from the painting area. This will dramatically reduce the chance of fire.

You may, if you do enough painting, need a spray painting booth or room. These are designed specifically for a spray painting operation. Contact your local fire department for the design requirements.

FIRE PROTECTION EQUIPMENT

You should store an extra 20B fire extinguisher near each flammable liquids storage area, room, or cabinet. Some operations may require sprinkler protection. Check with your local fire officials for code requirements.

Flammable Liquids Checklist

Date: _____   Name: _____

GENERAL STORAGE PRACTICES

_____ No Class I flammable liquids stored in polyethylene containers

_____ No flammable liquids stored in glass containers

_____ All flammable liquids stored in approved safety cans that have tight-fitting, self-closing lids and pressure-relief vents

_____ All fifty-gallon drums well-maintained and have tight-fitting lids and seals

_____ All drums properly cleaned before being filled with another flammable liquid

_____ All drums used for transferring flammable liquids grounded

_____ All flammable liquids tanks and containers equipped with automatic vents

_____ All flammable liquids containers labeled to DOT standards

_____ All flammable liquids used for cleaning stored in approved solvent tanks

_____ No waste flammable liquids dumped in ditch or sewer system or on the street

_____ All flammable liquids stored a safe distance from possible ignition sources

_____ All flammable liquids stored a safe distance from oxidizing or reactive chemicals

_____ All empty flammable liquids containers treated as full until cleaned properly

_____ All flammable liquids container lids and seals closed

_____ All flammable liquids stored a safe distance from exits

_____ All flammable liquids storage areas posted with warning signs

OUTDOOR STORAGE PRACTICES

_____ All flammable liquids stored a safe distance from the building

_____ Fire wall installed between flammable liquids and building wall

_____ All flammable liquids containers stored unstacked

_____ All storage areas built with a dike around flammable liquids

_____ All storage areas fenced for security

_____ Local fire department contacted for approval to abandon any large empty flammable liquids tanks

ALLOWABLE QUANTITIES

_____ Local fire authorities contacted for requirements specific to your business

Restaurants/Apartments/Hotels

_____ No unnecessary flammable liquids stored on premises

_____ All flammable liquids amounts over 10 gallons stored in a flammable liquids cabinet

Schools And Offices

_____ No unnecessary flammable liquids stored on premises

_____ All flammable liquids amounts over 10 gallons stored in a flammable liquids cabinet

_____ All flammable liquids cabinet storage capacities limited to no more than 120 gallons flammable liquids per cabinet

_____ All flammable liquids cabinet storage areas limited to no more than three cabinets stored per area

_____ Fire department contacted for storage requirements for larger amounts

Wholesale And Retail Stores/General Warehouses

_____ Fire department contacted for requirements specific to your business

_____ No unnecessary flammable liquids stored on premises

_____ All storage areas limited to a maximum of two gallons of flammable liquids per square foot of storage space

_____ All storage areas above the first floor limited to one gallon of flammable liquids stored per square foot of storage space

_____ No Class I-A flammable liquids stored in basement

_____ All flammable liquids containers stored near the ground

_____ All flammable liquids containers stored unstacked

_____ All flammable liquids storage racks secured

_____ All flammable liquids storage distributed over large area

Industrial Complexes

_____ Local fire department contacted for requirements specific to your business

Service Stations

_____ No Class I flammable liquids stored in a building with a basement or work pit unless adequate ventilation provided

_____ All welding areas equipped with extra ventilation

_____ All heating and electrical systems maintained in excellent condition

_____ "No Smoking" signs posted

_____ "Shut Off Engine" signs posted

_____ All pumps protected by concrete islands or protective posts

_____ Emergency shutoff switch installed

_____ Emergency shutoff switch labeled, clearly visible, and accessible

FLAMMABLE LIQUIDS CABINETS

_____ All metal flammable liquids storage cabinets built of five-one-hundredths-inch steel

_____ All metal flammable liquids storage cabinets built double-walled with an inch or two airspace between the walls

_____ All wood flammable liquids storage cabinets built at least one-inch thick

_____ All flammable liquids storage cabinet joints and corners built watertight — metal cabinets riveted or welded and wood cabinets secured in two directions

_____ All flammable liquids storage cabinets built with a two- or three-inch sill around the bottom

_____ All flammable liquids storage cabinets built with self-closing doors

_____ All flammable liquids storage cabinets limited to no more than 60 gallons of Class I flammable liquids per cabinet

_____ All flammable liquids storage cabinets limited to no more than 120 gallons of flammable liquids per cabinet

_____ All flammable liquids storage cabinets spaced at least 100 feet apart

_____ All flammable liquids storage areas limited to no more than three cabinets per room or area

SPRAY PAINTING

_____ All possible ignition sources eliminated

_____ "No Smoking" signs posted

_____ All heating and electrical systems maintained in good condition

_____ All electric lamps and heaters removed from area

_____ Spray painting room or booth used if necessary

FIRE PROTECTION EQUIPMENT

_____ All flammable liquids cabinets, storage areas or rooms equipped with an extra 20B fire extinguisher

_____ Sprinkler system installed if necessary

6.
Compressed Gases

 This section explains how to handle compressed gases. It includes general safety guidelines for the use and storage of flammable and non-flammable compressed gases.

Compressed gases can become dangerous in a hurry. Most are flammable and all are stored under pressure. If a container leaks, you will have instant trouble. First of all, flammable compressed gases leave their containers in vapor form. This vapor can ignite instantly with the smallest spark. Some compressed gases, especially oxygen, will add to the intensity of the fire. If a hot object is near an oxygen leak, it can start a nice big fire before you know it. In addition, if a portable cylinder springs a leak, it can turn into a wild torpedo trying to damage every expensive piece of equipment it can find.

Treat all compressed gas containers and piping with care and caution. Here are some suggestions for handling them.

GENERAL SAFETY PRACTICES

Several safety measures should be adopted when working with compressed gases. First of all, store the gases in the proper containers. All containers should be marked with the DOT stamp of approval. If the containers aren't marked, contact your local fire department for specific container requirements.

Compressed gas containers are either cylinders or tanks. Cylinders stand upright and contain less gas than tanks, which are mounted horizontally.

All tanks, cylinders, and pipelines should be protected from physical damage. It's also a good idea to store all your compressed gases in a secure room to prevent the untrained or mischievous from messing with the on/off control valves.

Finally, install each emergency shutoff valve in a visible and accessible place on the distribution line. Post a sign marking the location of each valve. Contact your local fire department for the requirements governing the number and location of these valves.

All cylinders should be secured so they don't fall over and spring a leak. Mount brackets to hold them or run a chain around all cylinders.

FLAMMABLE COMPRESSED GASES

Liquefied Petroleum Gas (LPG)
And Liquefied Natural Gas (LNG)

These two gases are highly flammable. Post "No Smoking" signs near every LPG or LNG container. No one should smoke or have any flame within 50 feet of any container. If you have an LPG or LNG tank outside the building, keep all trash, weeds, and brush at least 25 feet away. Hopefully, this distance will prevent the gas from igniting if a fire starts in the refuse or brush.

All tanks and piping should be protected from damage inflicted by vehicles. Protective concrete posts or curbs are good prevention devices to install. Any vehicle, even a slow-moving one, can cause enough damage to create a leak. Also, a hot motor can be a dandy ignition source.

In addition, cyclinders should be securely mounted to keep them from falling over. Use brackets or chain them together.

All containers should have at least one safety-relief valve. This valve allows gas to escape if too much pressure builds in the container. This increase in pressure will probably only occur during a fire, and the operation of the valve is crucial to keep the container from exploding. In a fire, the explosion of the container is a far greater hazard than the escaping gas. An explosion will significantly increase your damages.

Emergency shutoff valves should be installed and readily accessible. These valves should not be close to the cylinders or tanks because someone will have to get to them in a fire. Post a sign to mark the location of these valves.

Compressed Oxygen Gas

If you use oxygen equipment, keep all oils or petroleum products a safe distance from the oxygen piping. These chemicals will destroy the seals in the piping and create a gas leak. Oxygen will greatly increase the intensity of a fire. Post "No Smoking" signs near all oxygen equipment.

In addition, keep your electrical system in good condition, especially in areas near oxygen cylinders or equipment. A spark will really take off in an oxygen environment.

NON-FLAMMABLE COMPRESSED GASES

The primary concern with non-flammable gases is to prevent the cylinders from being damaged. This will help avoid the "torpedo" effect mentioned earlier, which can happen if the operating valve is knocked off. One way to avoid this problem is to securely mount all your cylinders. Attach a chain to the wall on one side of the cylinders, run it around them, and attach it to the far side. This will keep the cylinders in place.

Compressed Gases Checklist

Date: _____   Name: _____

GENERAL SAFETY PRACTICES

_____ All compressed gases stored in proper containers

_____ All compressed gas containers and pipelines protected from physical damage

_____ All compressed gas cylinders or tanks stored in a secure room or area

_____ All emergency shutoff valves installed in visible and accessible places

_____ All emergency shutoff valves posted with signs to mark their locations

_____ All compressed gas cylinders secured

FLAMMABLE COMPRESSED GASES

LPG or LNG Gases

_____ "No Smoking" signs posted near all LPG or LNG containers

_____ All LPG or LNG containers stored at least 50 feet from any flame or smoking area

_____ All LPG or LNG tanks stored at least 25 feet from any trash, weeds, brush, or other combustibles

_____ All LPG or LNG tanks and piping protected by posts or curbs to prevent possible damage

_____ All LPG or LNG cylinders mounted securely

_____ All LPG or LNG cylinders or tanks equipped
with at least one safety-relief valve

Compressed Oxygen Gas

_____ All oils or petroleum products stored a safe distance from oxygen equipment or piping

_____ "No Smoking" signs posted near all oxygen
cylinders and equipment

_____ All heating and electrical systems maintained in
good condition, especially near oxygen
equipment

NON-FLAMMABLE COMPRESSED GASES

_____ All non-flammable gas cylinders protected from
physical damage

_____ All non-flammable gas cylinders secured

7.
Portable Fire Extinguishers

 This section contains everything you've always wanted to know about fire extinguishers, and more. It explains how to classify fires and fire extinguishers, how to operate an extinguisher, how to choose the right extinguisher for your needs, when to have your extinguishers inspected, and how to maintain them.

If you don't have a sprinkler system, portable fire extinguishers are your first defense against fire. It is important that all your employees know where they are and how to use them. If the extinguishers are used promptly and properly, they can extinguish a fire that might otherwise do extensive damage.

CLASSIFICATION

Fires are divided into four classes, A, B, C, and D. Each relates to the burning material. Class A fires involve ordinary combustibles like paper, wood, cloth, and most plastics; Class B, flammable liquids; Class C, electrical equipment; and Class D, combustible metals like magnesium.

These classes require different forms of extinguishing agents. Each extinguisher is stamped with the letter of the class of fire it is designed to extinguish. Class A extinguishers usually contain water; Class B, dry chemical; Class C, carbon dioxide; and Class D, dry powder. Some extinguishing agents work on a variety of fires. You don't need to have a different extinguisher for each class of fire.

Fire extinguishers are also stamped with a number that relates how much fire they will put out. This number does not tell you exactly how much fire the extinguisher will put out, but indicates the amount it will work relative to the size of other extinguishers. For example, a 4A extinguisher will put out twice as much fire as a 2A extinguisher, at least theoretically.

This combination of numbers and letters gives the extinguisher its rating. The most common extinguisher is rated 2A10BC. This will work on combustibles, flammable liquids, and electrical fires. It contains a multi-purpose dry chemical extinguishing agent and is usually considered the minimum size required to effectively extinguish a fire. Some uses require different ratings. Check with your local fire department for minimum rating requirements for your business. You can purchase most extinguishers at hardware stores or fire prevention equipment companies.

Some fire extinguishers have been designated as obsolete. They have been shown to be dangerous, ineffective, or unreliable. If you have any foam, soda-acid, cartridge-operated, or vaporizing-liquid extinguishers, or any extinguishers that must be turned upside down to operate, throw them away. These extinguishers are dangerous to the operator because they conduct electricity, cannot be turned off, and are probably old enough to be unreliable. They are also corrosive to metals. One interesting point — vaporizing-liquid extinguishers contain carbon tetrachloride, which forms phosgene when it burns. The common name for this chemical during World War I was mustard gas. You definitely want to get rid of this one.

Check with your local fire department for a current list of obsolete extinguishers.

OPERATION

How the extinguishing agents work is a bit complex and boring. It is a chemistry class in itself. What is important though, is how to use an extinguisher. No matter what they look like or which agent they contain, all extinguishers operate in basically the same manner.

Most extinguishers have some sort of safety pin that must be pulled before it will work. The pin keeps the operating handle from being accidentally pushed. The pin is usually held in place by a small wire or plastic strap. Pull the strap and it will break easily. The extinguisher is now ready to use.

The next and most crucial operation is aiming the extinguisher. If you aim at the flames, it will not extinguish the fire. Aim the nozzle at the base of the fire. The fire starts at the base so this is where the extinguishing agent will do the most good. Hold the operating handle and squeeze. This discharges the extinguishing agent. Continue discharging the agent until the fire is completely out. Stand by for a minute or two; sometimes a fire will flare up again. You don't want this to go unnoticed.

Know how to operate your extinguishers before you need to use them. Every fire extinguisher should have operating instructions attached to it. These instructions should be facing out so they are visible. No matter what kind of extinguisher you have, read the operating instructions or invite your local fire department to visit once a year to show your employees how to operate the extinguishers.

LOCATION

Not only do you need to know how to operate the extinguishers but also where to find them. Your fire extinguishers should be in visible and accessible locations. When you need one, you usually need it in a hurry. You don't want to waste precious time searching for an extinguisher or moving furniture or storage to get to it. Fires grow quickly.

Mount your extinguishers in visible and accessible places such as on walls or posts. This will also keep them from being knocked over and damaged. Extinguishers are expensive to fix or replace. The top of the extinguisher should be no more than five feet above the ground. This allows shorter employees to lift an extinguisher off the hanger more easily.

If your extinguishers are still hard to see, use wall markings or signs to indicate their locations. The signs should be in contrasting colors and high enough on the wall to be clearly visible. In some warehouses, structural posts that hold extinguishers are painted red. Signs are available at your local fire department or most hardware stores and fire protection equipment companies. Occasionally, you may want to ask your employees if they know where to find the extinguishers. Remind them where the extinguishers are located and how to use them. Remember, the extinguishers are your first defense against fire.

Some extinguishers seem to disappear now and then. If you have a theft problem, install a system that sounds an alarm whenever the extinguishers are removed from their hangers or cabinets. Check with your local fire protection equipment company for other alternatives.

If a fire occurs, you have to decide whether it is best to use the extinguisher, or, if the fire is too big, to run! This is a difficult decision. I can only suggest that you be on the safe side and run. Call the fire department immediately. After all, fire-fighters are the ones who get paid to fight fire.

USE

The kind of extinguisher you should use depends on the hazards that exist in your business. If you work with common combustibles such as paper, wood, and plastic, use a pressurized water extinguisher. Most of these are rated as 2A extinguishers.

If your environment includes any quantities of flammable or combustible liquids such as gasoline, oil, or solvents, you need a dry chemical extinguisher with a minimum 10B rating.

Commercial kitchens with deep fat fryers, grills, and similar equipment need a 40BC extinguisher. This serves as a backup to any required fixed fire protection equipment.

Electrical equipment requires some special consideration. Water or dry chemical extinguishers can corrode or short circuit electrical equipment. The preferred extinguisher is carbon dioxide or Halon. Halon is especially good for sensitive equipment like computers. These extinguishers usually have a minimum rating of 10BC. The B indicates they can also be used on flammable liquids.

For general use, a dry chemical extinguisher with a minimum rating of 2A10BC is sufficient. This is also a good extinguisher for your home.

Check with your local fire department or fire protection equipment company for recommended extinguishers to serve your needs.

How many extinguishers you need depends on the size of the area you want to protect. A general rule of thumb is that you should only have to walk 75 feet to any fire extinguisher. This is the total travel distance, not necessarily the straight-line distance. If you have several storage racks or office partitions to negotiate, you need more extinguishers.

INSPECTION AND MAINTENANCE

Every fire extinguisher should be checked and inspected monthly. Note the date and person who performed the inspection on the inspection tag. All extinguishers should have one of these tags.

This monthly check should be done to ensure that all extinguishers are fully charged, visible, and accessible. In addition, the nozzle should be unobstructed, the safety seals in place, and all parts of the extinguisher free from damage.

The annual inspection includes loosening the extinguishing agent. To loosen the powder in a dry chemical extinguisher, turn the extinguisher upside down and shake it. Carbon dioxide extinguishers don't have a pressure gauge and should be weighed to make sure they are charged.

Your local fire department may not allow you to do these checks and inspections or, you may not want to perform them. In either case, a fire protection equipment company can do them for you. Check with your local fire department for regulations governing these checks and inspections.

Generally, only fire protection equipment companies are allowed to perform service maintenance and hydrostatic testing on your extinguishers. Hydrostatic testing is a pressure test of the tank. The tank is filled with water until nearly full and then pumped with air to test pressure. If the tank is faulty and ruptures, it will only spill water instead of exploding altogether. Extinguishers with stainless steel tanks should have a service maintenance and hydrostatic test every five years.

Portable Fire Extinguishers Checklist

Date: _____   Name: _____

CLASSIFICATION

_____ All extinguishers rated at least 2A10BC unless specific hazards dictate otherwise

_____ All obsolete extinguishers replaced with appropriately rated extinguisher

_____ Fire department contacted for current list of obsolete extinguishers

---

OPERATION

_____ All employees instructed about proper use of extinguishers

_____ All operating instructions visible on tank

LOCATION

_____ All extinguishers visible and accessible

_____ All extinguishers placed no higher than five feet above the ground

_____ All extinguishers protected with an anti-theft alarm system, if necessary

USE

_____ All pressurized water extinguishers used for common combustibles rated a minimum 2A

_____ All dry chemical extinguishers used for flammable or combustible liquids rated a minimum 10B

_____ All Halon or carbon dioxide extinguishers used for electrical equipment rated a minimum 10BC

_____ All dry chemical extinguishers used for general use rated a minimum 2A10BC

_____ All extinguishers located within a 75 feet travel distance of user

INSPECTION AND MAINTENANCE

_____ All extinguishers checked monthly

_____ All extinguishers inspected annually

_____ All service maintenance and hydrostatic tests scheduled and completed as follows:

Extinguisher	Service Maintenance	Hydro Test
Pressurized Water	5 Years	5 Years
Carbon Dioxide	5 Years	5 Years
Dry Chemical (Stainless)	5 Years	5 Years
Dry Chemical	6 Years	12 Years
Halon	6 Years	12 Years

8.
Sprinkler Systems

 This section explains the benefits of a sprinkler system and tells how to keep it up and running. It includes maintenance tips for the riser, supply lines, and sprinkler heads. It also sets forth an easy-to-follow schedule for inspecting and testing your system.

A sprinkler system is an excellent defense against fire. It can extinguish a fire before it gets out-of-hand. An Australian study showed sprinkler systems to be over 90 percent effective in extinguishing fires. If you have a sprinkler system, keep it operational so it can effectively do its job.

An improperly maintained system may not only fail to operate when it's needed, it may operate when it's NOT needed. Either way, it will cost you more than the maintenance. Proper maintenance is a good investment in your sprinkler system and your business.

The maintenance not only effects the operation of the system but also the ability of the fire department to control it. The firefighters may want to add water to the system to control the fire, or they may want to shut it down to minimize the water damage. The basics of sprinkler system maintenance are the same regardless of which system you own.

To understand the "how and why" of maintenance, it helps to understand the basic operating principles of a sprinkler system. The system receives water from a main supply pipe that runs underground from the public water main. This pipe comes into your building at a point called the riser. The riser is the control area for the system. A variety of valves and piping distribute water to gauges, alarm systems, and other incidental systems.

The main shutoff valve is also located at the riser, but it usually operates from the outside wall of the building or from a valve located along the water supply line. A shutoff valve on the outside of the building is usually an outside stem and yoke (OS&Y) valve. A shutoff valve along the supply line is usually a post indicator valve. Either valve does the same job; it serves as the on/off switch for the water supply and sprinkler system. Post indicator valves can be located in a variety of places. Make sure you know where they are located.

Another valve near the riser, which is called the clapper valve, keeps water from continuing up through the system until it is needed. This valve has a latch that releases when a sprinkler head opens. Water will then flow freely to the sprinkler heads.

A feeder line begins at the riser and continues up to the branch lines. The branch lines hold the sprinkler heads. These feeder lines and branch lines are held by hangers. Bracing is installed at every bend in the system for additional strength.

All sprinkler heads are designed to open at a certain temperature and release water onto a fire. Each head has a cap over the opening. This cap is held in place through tension. When the temperature gets high enough, the tension is released and the cap comes off. Water is then discharged through the opening. A deflector on the end of each sprinkler head turns the water into a spray. This spray is more effective than a stream in extinguishing a fire.

If your system has over 100 sprinkler heads, you may be required to connect it with an alarm monitoring station. This station must be attended 24 hours a day, seven days a week. This requirement gives the fire department an extra jump on a fire. Check with your local fire department for specific requirements.

Each part of the sprinkler system requires some sort of maintenance. Let's examine the maintenance guidelines for each component of the system.

RISER MAINTENANCE

All of the control valves must be in their proper positions. Some are open; some are closed. Contact your fire department or fire protection equipment company; it will know the correct position for each valve.

The riser should be clear of any obstructions. Also, don't place anything nearby that could fall and damage it. Some of the valves and piping are small and can be easily damaged. Allow three feet of clearance around the riser.

The riser has a main shutoff valve along the supply line or on the outside wall of the building. The firefighters will want to get to this control valve in a hurry to limit water damage. Keep this valve visible and accessible. Allow three feet of clearance around the shutoff valve.

Secure the shutoff valves in the open position. Put a chain with a lock through the shutoff handles to keep the system from being accidentally or maliciously turned off. Another solution is to install an electronic tamper switch. This will send a signal to an alarm station if anyone, including the neighborhood trouble-maker, tries to close the valves. If the system is off when you need it, you'll have to reach deeper into your pocketbook.

Properly label all the valves at the riser. This will save time when the firefighters need to control the system.

Your system also has a supply valve that allows the fire department to pump additional water through your system. This valve is called the sprinkler connection. It may be near your building or in the parking lot, depending on where the fire department wanted it installed when the building was constructed. It should be visible and easily accessible. Allow three feet of clearance around the sprinkler connection.

SUPPLY LINE MAINTENANCE

The system's supply lines, which are called the feeder and branch lines, are hung from the ceiling by hangers. All of these hangers should be securely fastened to the ceiling. If they are loose, the lines may start to sag and crack. Your system will then leak like a sieve.

Each bend in the supply line piping is usually required to have a brace. This prevents movement in the system during its operation or an earthquake. How much bracing is required depends on the risk and magnitude of earthquakes in your area. Contact your fire protection equipment company; it will know how much bracing is necessary. Keep the bracing securely fastened; it will help prevent leaks.

Keep all stock or merchandise at least three feet from the supply lines, and don't attach or hang anything from the supply line piping.

You have additional maintenance concerns if you live in a cold climate. Set cold weather valves in the proper position if the temperature drops below freezing. If your system doesn't have this kind of valve, set your thermostats above 40 degrees Fahrenheit or four degrees Celsius. This will keep the system from freezing. Freezing takes the system out of action and may cause some terrific leaks.

SPRINKLER HEAD MAINTENANCE

All sprinkler heads should be in good condition. Corrosion on the heads may mean they will spring a leak. Paint on the heads will change the operating temperature and may mean they will operate too early or, too late.

If the heads are located where they can be hit by something, install protective guards. These guards protect the sprinkler heads from being damaged, which can cause accidental discharges. All sprinkler heads should have a three-foot clearance from any stock, merchandise, or other obstructions. They need space to operate properly.

If you add any decks or partitions to a building already equipped with a sprinkler system, extend the system to the new areas. This maintains the integrity of the system. If the new area is not equipped with sprinkler heads and it catches on fire, the fire may grow too large for the rest of the system to extinguish.

Keep a supply of extra sprinkler heads on hand. If some of your heads don't pass inspection and you don't have extras, the fire department can close your business until new ones are purchased and installed. You will also have to replace the sprinkler heads if you have a fire. If you don't have extra heads, you may have to stand fire watch until the old ones can be replaced.

Either situation is a needless waste of your time and money. Also, make sure all extra sprinkler heads have deflector plates and the same temperature rating as the heads currently in place.

Here's a rule of thumb to follow for how many extra heads to keep on hand.

Number of Heads in System	Number of Extra Heads
less than 300	6
300 to 1,000	12
over 1,000	24

SPECIAL TESTING

Contact a fire protection equipment company to inspect and service your sprinkler system every five years. However, systems that protect commercial kitchen hood and duct arrangements should be serviced more frequently. Because the accumulation of grease in these areas creates a fire hazard, these systems should be serviced every six months or after activation. If your system was activated, contact the health department to supervise the clean up.

If your system has alarm or automatic notification devices, have them tested once a year. These devices are important because they give the fire department an early warning of fire. If the devices don't work, or work when you don't want them to, it can cost you dearly. Be aware that many fire departments have started to fine businesses for repeated false alarms. In either case, it pays to have your alarm system in good condition.

If your system is being tested or shut down for repairs, notify your fire department. The firefighters need to know because it will change their tactics if you have a fire.

Sprinkler Systems Checklist

Date: _____   Name: _____

RISER MAINTENANCE

_____ All control valves set in proper positions

_____ Three-foot clearance zone maintained around riser

_____ Three-foot clearance zone maintained around main shutoff valve

_____ Main shutoff valve secured in open position

_____ All valves labeled properly

_____ Three-foot clearance zone maintained around sprinkler connection

_____ Sprinkler connection visible and accessible

SUPPLY LINE MAINTENANCE

_____ All hangers fastened securely to ceiling

_____ All bracing fastened securely

_____ All supply lines unobstructed and unencumbered

_____ Three-foot clearance zone maintained around supply lines

_____ All cold weather valves set in proper position

_____ Thermostats set above 40 degrees Fahrenheit or four degrees Celsius

SPRINKLER HEAD MAINTENANCE

_____ All sprinkler heads maintained in good condition

_____ All sprinkler heads unpainted and uncorroded

_____ Protective guards installed around all sprinkler heads, where necessary

_____ Three-foot clearance zone maintained around all sprinkler heads

_____ All areas of the building equipped with sprinkler heads

_____ Extra sprinkler heads stored on premises as follows:

Number of Heads in System	Number of Extra Heads
less than 300	6
300 to 1,000	12
over 1,000	24

_____ All sprinkler heads equipped with deflector plates

SPECIAL TESTING

_____ Sprinkler system inspected and serviced every five years

_____ Commercial kitchen hood and duct sprinkler systems inspected and serviced every six months or after activation

_____ Alarm or automatic notification devices inspected and tested every year

_____ Fire department notified when sprinkler system is off

_____ Monitoring station installed for sprinkler systems with over 100 sprinkler heads

9.
Exits

EXIT If a fire does start, how do you get out of the
building safely? This section explains how to design and maintain your escape routes — exit paths
and exit doors.

Exits are crucial for the safe and speedy departure of the
building's occupants. They are critically important in an
emergency. Many people have been killed in fires or other
disasters because they could not get out of a building fast
enough. All buildings, big or small, need proper exits and
pathways.

Every building is required to have exits and exit pathways.
The number and arrangement of these exits varies. These
factors depend on the size of the building, its purpose, and its
fire protection systems. For example, a large office building
requires more exits than a large warehouse because it has a
higher people population. More people need more exits. A
small spray painting operation needs more exits than a small
office because spray painting is more hazardous. A building
with a sprinkler system requires fewer exits than an identical
building without one, because the system will stop or slow the
spread of fire and give the employees more time to evacuate.

No matter how many exits your building has, if they are not
maintained properly, you could find yourself helplessly unable
to escape a very hazardous situation. If you neglect your exits
and something goes wrong, someone may want to see you in
court. Here are some safety points about exits.

EXIT PATHS

Do not obstruct any exit path. Give your employees a clear
path to the exit door.

Clearly define your exit paths. Provide easily followed paths at least four feet wide. Aisles between storage or shelving should be at least three feet wide. The exit paths should not pass through kitchens or any other functional room.

Stairwells can be vital escape routes. Don't store any combustible material in them. It's dangerous to have a potential fuel source in an exit.

If the stairwell goes below the street level, place a sign on the street level exit so people don't pass it by in the confusion of an emergency.

Every exit should open to an exit passage or sidewalk. This prevents people from exiting into another part of the building or into the business next door. You want your people to get outside the building. This is the safest place when there is a fire inside.

If exiting is the least bit complicated, post exit maps in several convenient and visible locations so your employees and customers can easily find the nearest exit. These exit directories should indicate the person's location relative to the exits. Use a contrasting color to indicate the path to the exits. These maps are especially handy when you are on an upper floor and the elevators have been shut down.

Do not attempt to use the elevators during a fire. They may shut down at any time and you will be stuck. If you work in a high-rise building, know the location of the exit stairwells. Some buildings also have "safe" areas where you can wait out the fire if you can't escape the building. Find out if your building has these areas and where they are located.

Long hallways, such as those found in apartments, should have sufficient lighting to get people safely to an exit. Operating in the dark will scare the daylights out of anyone, especially in an emergency. In addition, you should have lighted exit signs along the hallways to steer everyone to the exit. Most hotels and apartments with interior hallways are required to have lighted exit signs.

EXIT DOORS

You should be able to open all exit doors easily and quickly. When you need out, you need out NOW! If you have to fumble around with keys and latches you'll lose precious time.

A door with "panic" hardware is the easiest to open. This hardware is a simple bar across the door that you push to open the latch. All you have to do is bump into the bar to open the door. In an emergency, a door with this kind of hardware is the easiest to operate.

Exit doors should be clearly distinguishable from the surrounding wall. This avoids confusion. For example, some glass doors are hard to distinguish from the windows around them. Place a sign above the exit if necessary.

Exit doors should swing out. When leaving a building, it is easier to push the door out than to pull it in, especially in an emergency.

Most stairwells are required to have self-closing doors. Self-closing doors should remain closed at all times. These doors prevent fire from traveling up or down the stairwell. Don't obstruct these doors or prop them open.

Automatic-closing doors are designed to stay open until a fire occurs. This type of door is usually present in long hallways. Don't obstruct or otherwise mess with these doors. They usually have a fusible link that breaks in a fire and allows the doors to close. This link must be kept in good condition and shouldn't be replaced with something that keeps the door open.

Illuminate your exits if the building is open after dark. Let people see the exit. A small light that illuminates the sign above the exit will probably be sufficient.

Your exit doors should be spaced a reasonable distance apart; otherwise, all your exits can be blocked by a small fire. Don't put all your eggs in one basket. If you have a choice, put exit doors on opposite sides of the building. Check with your local fire department for more design help.

Exits Checklist

Date: _____   Name: _____

EXIT PATHS

_____ All exit paths unobstructed

_____ All exit paths defined clearly

_____ All exit paths and aisles maintained at least four feet wide

_____ All aisles maintained at least three feet wide between storage or shelving

_____ No combustibles stored in stairwells

_____ All street level doors posted with exit signs, especially on doors in stairwells that go below street level

_____ All exits designed to open into an exit passage or outdoor sidewalk

_____ Exit maps or directories posted in visible places

_____ All employees instructed about the location of exits

_____ All employees instructed about the location of "safe" areas

_____ All long hallways equipped with sufficient lighting and direction signs

EXIT DOORS

_____ All exit doors unobstructed and open easily

_____ All exit doors distinguished easily from sur-
roundings

_____ All exit doors designed to swing in the direction
of exiting — out

_____ All self-closing or automatic-closing doors unobstructed and operate properly

_____ All exits visible at all times and lighted after dark

_____ All exits spaced evenly throughout the building

10.
What To Do If You Have A Fire

Although the purpose of this book is to prevent fires, it's still possible that something may go wrong. Fires are caused by people because people are not perfect. Even if you've taken all the steps to prevent one, you may have a fire someday. Plan what to do if you have one; there is always a chance that Murphy's Law will knock on your door.

SOUND THE ALARM

If a fire occurs in your business, it is critical to notify everyone so they can get out of the building safely. If you don't have a fire detection and alarm system, you may want to install a loud horn that can be set off by anyone who detects a fire. This can be as simple as placing air horns in locations where they will be heard by everyone. You can place them near your fire extinguishers or install smoke detectors. Check with your local fire department or fire alarm company for more ideas.

It doesn't matter what kind of alarm system you have as long as everyone knows what to do when they hear the signal for fire.

EVACUATE

Some people hear fire alarms time after time and start ignoring them if they have always been false. People get annoyed at these frequent false alarms, especially during busy work days or if they are on the 35th floor and are supposed to leave the building. These are the people who get hurt or killed when the real one hits. Remember the boy who cried "wolf" too many times.

It seems much better to be alive and annoyed than dead and buried. If the alarm system is not working properly, get it fixed. Every false alarm costs you money. When the alarm does go off, it is extremely important that everyone takes it as the real thing. It just might be.

If the real thing does happen and the alarm sounds, everyone needs to know how to leave the building quickly. It's the smoke that kills people, and it travels fast. People won't be able to find their way out if the exits are blocked or the building is so large they get lost. You may want to review Section Nine on exits. Keep the exits clear and post exit maps if your building is a big one.

If you see or smell smoke, stay low. Crawl if you have to; there will be less smoke close to the ground. Check all doors with the back of your hand. It it's hot, don't go in. Use an alternative route. Everyone should know at least two escape routes.

You and your employees can help stop the fire from spreading if you close all the doors on your way out. The fire will then have to burn through the doors before it can race into the next room. Closing the doors can save a lot of damage costs, but it is more important to evacuate the building and save yourself. Close the doors if you can, but don't lock them. Someone may still need to get out and the fire department will want to get in.

The next important step is to notify the fire department. If you have a fire detection and alarm system, this is probably done for you. If you don't have such a device, someone will have to call the fire department. If the situation is at all hazardous, don't let anyone stay inside the building to call. Get everyone out first and then worry about finding a phone. There isn't a building on earth that is worth one life.

Count heads to make sure everyone is out of the building. This should be easy if you only have a few employees. If you have more, you may want to have supervisors account for certain people. Break the number into manageable groups. However you do it, count off and make sure everyone is out. Let the firefighters know if anyone is missing. They will want to know so they can prioritize their actions.

Once everyone is out of the building and the fire department is on the way, do not — under any circumstances — let anyone back into the building. Photographs on the desk or the favorite sport coat on the chair will have to be left behind. Many people have died because they went back into a burning building. If you get out safely once, don't push your luck. Stay out and don't let anyone else back in. The firefighters get paid to go in. Let them rescue your personal effects. They will do their best.

HELP THE FIRE DEPARTMENT

The next thing to do is to find out if anyone knows what happened — who saw the fire, where it is, and that sort of thing. This will be helpful to the fire department because the firefighters won't have to search everywhere for the fire. The smoke will probably get dense enough to cut their visibility down to zero. Fighting fire is a lot like trying to find a time bomb in a pitch black room. You know it's there somewhere, but you don't know when it will go off. The firefighters have to find the fire before it finds them.

It is also important to let the firefighters know if there are any hazardous areas in the building. This may effect how they approach fighting the fire. Any details you can give them will be appreciated and may also help reduce the damage.

Even if you do nothing to prevent a fire, be prepared if you have one. It may save lives and reduce damage.

SUMMARY

- Plan ahead
- Notify everyone of the problem
- Exit calmly but quickly
- Close all doors if safety permits — do not lock them
- Notify the fire department
- Count heads
- Keep everyone outside the building
- Find out what happened
- Give details to the firefighters

11.
How To Survive A Fire Department Inspection

 When the fire department comes to visit, the most important thing for you to have is a positive attitude — toward the firefighters and the inspection itself.

Firefighters don't view inspections as the highlight of their day. They would rather be fighting a fire; that's where the excitement is. They may see inspections as a pain-in-the-rear responsibility that their administration has mandated. Besides, they are probably a little bit afraid of you, too.

If you butt heads with them or make them feel uncomfortable, they are likely to be defensive. They just may write you up for every little infraction to even the score.

Your most important objective then, is to make them feel comfortable. Let the firefighters know you appreciate their expertise and willingness to help you find and correct fire or life safety problems. It will make the whole inspection process easier if you demonstrate cooperation and a positive attitude.

When the firefighters arrive at your doorstep, greet them with a smile and a handshake. Try to remember their first names. They should tell you why they have come and ask if this is a convenient time for the inspection. If you really are too busy, explain the situation and see if you can arrive at a mutually acceptable date and time for the inspection. Once you make this appointment, keep it. If you repeatedly make and break appointments with them, they may think you have something to hide.

During the inspection, the firefighters will probably want information such as the names and phone numbers of people to call in case of an emergency.

This is a good time to tell them about your own fire prevention program. Tell them that you have read this book and have taken steps to become more aware of the hazards and more involved in fire prevention. Describe your own prevention and inspection program. Relate some of the problems you've discovered and explain the steps you've taken to correct them. This will show the firefighters that you really are interested and involved. It will also tell them that they shouldn't expect to find too many problems. In addition, you may want to pull out past fire department inspection records and offer them as a gesture of your willingness to cooperate.

When the actual inspection begins, the firefighters will want to look in every nook and cranny of your business. If you need to let someone know you are coming, do so before you set off on the tour. Make sure you have everything you need, such as keys, to complete the tour.

The inspectors — firefighters who are not chasing fires at that moment! — may want someone to accompany them. If you go with them, it will be a good opportunity for both of you. You will get a chance to explain how your business works, and they will be able to point out potential problems and explain how to solve them. This will start a conversation and will also give the firefighters a better understanding of your business. They may have to return under very unfavorable circumstances and the more they know about your business, the better they will be able to handle an emergency.

If you can't go on the inspection tour yourself, send someone who is very familiar with your operation. The inspectors may need a guide, and they may have questions that need a knowledgeable person to answer. You don't want the inspectors to get lost, and you don't want them to get bad information. If they have an escort, you can save yourself and the inspectors time and frustration.

If the inspectors find a problem or code violation, ask them to explain it to you. You need to fully understand what each problem is, why it is a problem, and how to correct it. If inspectors can't explain it, ask them to get back to you with the answers. At this point you may run into problems. Many inspectors know what the problem or violation is but don't fully understand why. If they can't explain it, don't jump on them. Don't make the inspectors look stupid, even if they are a little uninformed. They really can make life more miserable for you than you can for them.

Ask the inspectors who you can call to get a better understanding of the problem and how to correct it. It pays to know who to turn to for help. If you need answers, call some of the people listed in the Resources Section.

After the inspectors have examined your business and explained any problems, thank them for coming. This is just common courtesy. They will probably go outside and heave a big sigh of relief.

There is a saying in the fire service, "Telephone, telegraph, tell-a-fireman." If you make a positive impression on the people who come to inspect your business, the entire fire department will soon know you are a nice person and their friend. You will have started a positive relationship that will help you in the future.

SUMMARY

- Have a positive attitude
- Welcome the firefighters into your business
- Express your appreciation for their help and concern
- Be prepared
- Describe your own prevention and inspection program
- Give the inspectors past fire department inspection records
- Accompany the inspectors on the inspection tour
- Explain how your business operates
- Ask the inspectors to explain any problems
- Thank the inspectors for their help

PART THREE
HOW TO CREATE YOUR OWN
SUCCESSFUL PROGRAM

1.
Teamwork

Now that you've read the information in this book, what can you do with it? You can do a lot. A fire prevention program can be a starting point for an overall management philosophy.

Many of today's popular management books express how important it is for business owners to foster a team attitude between employees and management. This attitude promotes a feeling of mutual concern. The experts encourage you to actively involve your employees in the operation of the business. They also suggest you demonstrate respect for your employees and create a secure and comfortable atmosphere in their workplace.

All of these modern management techniques are tools to make your business a profitable, efficient, happy, and safe place to work. A fire and life safety program can help you attain these goals. Let's look at each of these techniques.

You can foster this team attitude and sense of mutual concern by starting a fire and life safety program and asking your employees to run it. Explain why you are starting the program. Tell them you want to make sure the business doesn't burn down; if it does burn down, they will be out of work and won't be able to pay the bills. And neither will you. You and your employees are on very common ground here. It is in this sense of mutual concern that you want to work as a team to prevent a disaster.

Also tell them why you want them to run it. Let them know you realize they are in the best position to find the problems. They work around them everyday. Give your employees complete responsibility for finding the potential problems. They find them; you fix them. This program can help the whole business operate as a team for the mutual benefit of all.

As a sideline to this team approach to fire and life safety, ask your people to bring up any other problems they notice in the business. Let everyone become involved in finding better ways to operate the business. You may be surprised by your employees' innovation and creativity.

If teamwork can be used in a fire and life safety program, it can be used in another positive way. Use it to let your employees contribute to their own welfare. This involvement can help your business prosper, and if the business prospers, so will its employees.

Even though this may create a feeling of financial security, your employees still need to feel safe on the job. Many employees are unhappy when they feel the workplace is a hazard to their very survival. So would you. The feeling of physical security is a basic human need. Give it to them.

A fire and life safety program will help them understand that you are concerned about them in more human terms. If they are responsible for this program, they will know the workplace is safe.

The key ingredient in all of these management techniques is the demonstration of your respect for each individual. This is what will make the program work for you.

All of these techniques try to pull the best qualities out of everyone. As Kenneth Blanchard, Ph.D., and Spencer Johnson, M.D., explain in their book, *The One-Minute Manager,*

"Help people reach their full potential. Catch them doing something right."

Reinforce good work. When someone does something right, note it and compliment it. That person will remember what he or she did and will be more likely to do something right again.

Anything that comes out of a fire and life safety program can be counted as something right. Every time a problem is found, whether it could have caused a minor inconvenience or a major disaster, you get an opportunity to reinforce good work. Being noticed for good work can really raise morale. Look for the good things your people accomplish. It will breed more good work.

A fire and life safety program will also give you an opportunity to do things right. As morale goes up so does efficiency and productivity. This means your business will be more profitable.

And all this will happen because of you. You will get all those warm and fuzzy feelings right along with your people. Everyone will look at the operation of the business in a little different light.

Armed with the information in this book, you can help yourself, your employees, and your business down the path to new profit records. As you can see, a fire and life safety program can do more for you than it may initially appear. Here are some ideas about how to start it.

2.

Organization

First of all, decide who will be responsible for the program. It may be you or someone you designate. Give that person this book to read. Your coordinator needs to understand the basics of fire prevention and life safety.

Meet with your coordinator to discuss the goals and operation of the program. Part Two of this book is broken into 11 sections. You may decide to go through one section each month or complete all the sections in a few days. This is a chance for input from the person or persons responsible. If it's their show, let them run it. The point is this: set a goal to complete a thorough inspection tour of your business. Your timeframe will depend on the size, configuration, and time constraints of the business.

Each of the first nine sections in Part Two includes a checklist. Use a copy of each checklist as a guide to help you find problems. After each inspection, have the inspector sign and date the checklist. If a problem is found, the inspector should describe the problem on a separate sheet of paper and note what corrective action was taken. Remember, just knowing what the problems are and where they exist can prevent a major disaster. You may use the Inspection Form included in the Appendix, or draft your own. Keep these completed inspection forms and checklists handy for the fire department. The inspectors will be impressed by your efforts.

To keep the program on track, set up regular meetings. Talk about any problems and decide how to deal with them. In addition, be open to discuss any operational problems or suggestions that may be good for the welfare of the business.

This program can be run as many different ways as there are businesses. Use your imagination, or better yet, let your employees use their imagination. Your business will be better for it all the way around. Also, remember to contact your insurance agent to see if you can qualify for a premium reduction by following the safety guidelines in this book or by implementing a fire and life safety program.

The most important point about the program is this: *START IT NOW.* Take a few minutes now to get it underway. It may be the best investment you ever make.

APPENDIX

Inspection Form

Date:

Inspector:

Personnel Notified:

Location and Description of Problem:

Recommended Solution:

Location and Description of Problem:

Recommended Solution:

Resources

It is up to your coordinator to locate the names and phone numbers of your local officials. Most of them can be found in the phone book. For easy reference, this section lists the resources by the function each performs.

FIRE DEPARTMENT

Fire Chief:
Telephone:

Fire Marshall:
Telephone:

Fire Inspector:
Telephone:

BUILDING DEPARTMENT

Building Official:
Telephone:

Building Inspector:
Telephone:

Electrical Inspector:
Telephone:

FIRE PROTECTION EQUIPMENT COMPANY

Company:
Representative:
Telephone:

Company:
Representative:
Telephone:

Company:
Representative:
Telephone:

Company:
Representative:
Telephone:

UNDERWRITERS LABORATORY

This is a testing lab. These people have a wealth of information on the fire safety of various products, appliances, and machinery.

Local Representative:
Telephone:

National Headquarters:
Telephone:

FACTORY MUTUAL

This is a another testing lab.

Local Representative:
Telephone:

National Headquarters:
Telephone:

INSURANCE COMPANY

Company:
Policy Number:

Local Agent:
Telephone:

Claims Representative:
Telephone:

MISCELLANEOUS RESOURCES

Add to your list as necessary.

Name:
Organization:
Telephone:

Name:
Organization:
Telephone:

Name:
Organization:
Telephone:

Name:
Organization:
Telephone:

About The Author

Robert Pessemier is a ten-year veteran of the fire service. He is a lieutenant with the Kent Fire Department near Seattle. Lt. Pessemier has been involved in fire prevention for all of his seven years with Kent, including a two-year assignment to the Fire Prevention Division. During this assignment, he gained valuable experience in fire prevention and life safety and also studied at the National Fire Academy in Maryland. His education also includes a bachelor of science degree from the University of Washington.

Lt. Pessemier is a recipient of the distinguished *Meritorious Service Award* for "courage and bravery beyond the call of duty."

INDEX

Acetone, 67

Acrylic Paint, 67

Active Prevention — *see* Fire
Prevention

Adapters — *see* individual entries

Aerosol Cans, 73

Air Conditioning Systems, 39

Aisles, 57, 60, 106, 108

Alarm Systems — *see* Fire Detection
And Alarm Systems

Allowable Quantities — *see*
Flammable Liquids

America, 15, 25, 27, 29, 33

Analogies, 22, 27, 29, 45

Apartments, 72, 78, 106

Appendix, 125, 127-143

Appliances — *see* Electrical And
Heating Appliances

Arcing, 45, 46, 49, 50

Ash Trays, 38, 41

Ashes, 37, 38, 41, 62, 63

Attics, 56, 59

Attitudes — *see* Business

Australia, 21, 97

Author, About The, 132

Bah-humbug, 21

Basements, 72, 73, 79, *see also*
Work Pit

Benzene, 67

Blanchard, Kenneth, Ph.D., 122

Blower, Explosion-Proof, 73

Boiling Point — *see* Flammable
Liquids

Branch Lines, 98, 99

Building
Code, 72
Construction, 37, 40, 43

Department, 129
see also Offices

Bunker Fuel, 68

Burke, Edmund, 17

Business
Attitudes and Misconceptions,
21, 22, 24, 34
Community, 21-25, 33
Fires, 45, *see also* individual
entries
Inspections, 115-117, 125
Interruption, 23, 45, 46, 50
Owners, 15, 27, 33, 34, 121
Productivity, 18, 122
Profits, 16, 25, 46, 122, 123

Canada, 17, 21, 28

Candles, 38, 41

Carbon Tetrachloride, 90

Cars, 50

Cartridge-Operated Fire
Extinguishers, 90

Ceilings, 40, 43, 48, 49, 50, 52, 56,
59, 99, 102

Celsius, 100, 102

Checklists, 18, 125
Commodities In Storage, 59-60
Compressed Gases, 86-87
Electrical Distribution Systems,
51-53
Exits, 108-109
Flammable Liquids, 76-81
General Considerations, 41-43
Portable Fire Extinguishers,
94-95
Sprinkler Systems, 102-103
Waste Materials, 63

UP IN SMOKE
Order Form

Phoenix Publishing
P.O. Box ~~3546~~ *15164*
~~Redmond~~, WA ~~98073-3546~~ USA
Seattle *98115-0164*

Satisfaction guaranteed or your money back!

_____ Yes! Please send me _____ copies of *Up In Smoke*

_____ Yes! Please contact me about special volume dis-
counts for orders of 25 copies or more.

I'm interested in _____ copies of *Up In Smoke*

Name: _____

Occupation: _____

Address: _____

_____ Zip: _____

Amount Enclosed

_____ copies @ $7.95 each   _____

Sales Tax: Washington residents add 7.9%
(63¢ per book)   _____

Shipping: $1 for the first book, 50¢ each
additional book
Delivery: 3-4 weeks   _____

Air Mail: $3 per book   _____

TOTAL AMOUNT ENCLOSED $_____